D0041403

Sa...
Library

Santiago Canyon College
Library

The Art of the Infinite

The Tower of Mathematics is the Tower of Babel inverted: its voices grow more coherent as it rises. The image of it is based on Pieter Brueghel's "Little Tower of Babel" (1554).

The Art of the Infinite:

The Pleasures of Mathematics

QA
295
.A85
2003

ROBERT KAPLAN AND ELLEN KAPLAN

Illustrations by Ellen Kaplan

OXFORD

UNIVERSITY PRESS

ocm 51962641

2003

Santiago Canyon College
Library

OXFORD
UNIVERSITY PRESS

Oxford   New York
Auckland   Bangkok   Buenos Aires   Cape Town   Chennai
Dar es Salaam   Delhi   Hong Kong   Istanbul   Karachi   Kolkata
Kuala Lumpur   Madrid   Melbourne   Mexico City   Mumbai   Nairobi
São Paulo   Shanghai   Taipei   Tokyo   Toronto

Copyright © 2003 by Robert Kaplan and Ellen Kaplan

Published by Oxford University Press, Inc.
198 Madison Avenue, New York, New York 10016

www.oup-usa.org

Oxford is a registered trademark of Oxford University Press

All rights reserved. No part of this publication may be reproduced,
stored in a retrieval system, or transmitted, in any form or by any means,
electronic, mechanical, photocopying, recording, or otherwise,
without the prior permission of Oxford University Press.

Library of Congress Cataloging-in-Publication Data
is available

ISBN-0-19-514743-X

Also by Robert Kaplan
The Nothing That Is: A Natural History of Zero

1 3 5 7 9 8 6 4 2

Printed in the United States of America
on acid-free paper

For Michael, Jane, and Felix

Contents

Acknowledgments   ix

An Invitation   1

Chapter One
Time and the Mind   3

Chapter Two
How Do We Hold These Truths?   29

Chapter Three
Designs on a Locked Chest   56

Interlude
The Infinite and the Indefinite   75

Chapter Four
Skipping Stones   77

Chapter Five
Euclid Alone   100

Interlude
Longing and the Infinite   131

Chapter Six
The Eagle of Algebra   133

Chapter Seven
Into the Highlands   167

Interlude
The Infinite and the Unknown   200

Chapter Eight
Back of Beyond   202

Chapter Nine
The Abyss   228

Appendix   263

Bibliography   315

Index   317

Acknowledgments

We have been unusually fortunate in our readers, who from four different perspectives brought our book into focus. Jean Jones, Barry Mazur, John Stillwell, and Jim Tanton put a quantity of time and quality of thought into their comments, which made the obscure transparent and the crooked straight. We are very grateful.

The community of mathematicians is more generous than most. Our thanks to all who have helped, with special thanks to Andrew Ranicki and Paddy Patterson.

No one could ask for better people to work with than Eric Simonoff and Cullen Stanley of Janklow & Nesbit, who make the gears that turn writing into reading mesh with ease; nor a better, more thoughtful editor than Peter Ginna, in whom all the best senses of wit unite.

The Art of the Infinite

∞

An Invitation

Less than All cannot satisfy Man.

—William Blake

We commonly think of ourselves as little and lost in the infinite stretches of time and space, so that it comes as a shock when the French poet Baudelaire speaks of "cradling our infinite on the finite seas." Really? Is it ourself, our mind or spirit, that is infinity's proper home? Or might the infinite be neither out there nor in here but only in language, a pretty conceit of poetry?

We are the language makers, and what we express always refers to something—though not, perhaps, to what we first thought it did. Talk of the infinite naturally belongs to that old, young, ageless conversation about number and shape which is mathematics: a conversation most of us overhear rather than partake in, put off by its haughty abstraction. Mathematics promises certainty—but at the cost, it seems, of passion. Its initiates speak of playfulness and freedom, but all we come up against in school are boredom and fear, wedged between iron rules memorized without reason.

Why hasn't mathematics the gentle touches a novelist uses to lure the reader into his imagination? Why do we no longer find problems like this, concocted by Mahāvīrā in ninth-century India:

> One night, in a month of the spring season, a certain young lady was lovingly happy with her husband in a big mansion, white as the moon, set in a pleasure garden with trees bent down with flowers and fruits, and resonant with the sweet sounds of parrots, cuckoos and bees which were all intoxicated with the honey of the flowers. Then, on a love-quarrel arising between husband and wife, her pearl necklace was broken. One third of the pearls were collected by the maid-servant, one sixth fell on the bed—then half of

1

what remained and half of what remained thereafter and again one half of what remained thereafter and so on, six times in all, fell scattered everywhere. 1,161 pearls were still left on the string; how many pearls had there been in the necklace?

Talking mostly to each other or themselves, mathematicians have developed a code that is hard to crack. Its symbols store worlds of meaning for them, its sleek equations leap continents and centuries. But these sparks can jump to everyone, because each of us has a mind built to grasp the structure of things. Anyone who can read and speak (which are awesomely abstract undertakings) can come to delight in the works of mathematical art, which are among our kind's greatest glories.

The way in is to begin at the beginning and move conversationally along. Eccentric, lovable, laughable, base, and noble mathematicians will keep us company. Each equation in a book, Stephen Hawking once remarked, loses half the potential readership. Our aim here, however, is to let equations—those balances struck between two ways of looking—grow organically from what they look at.

Many small things estrange math from its proper audience. One is the remoteness of its machine-made diagrams. These reinforce the mistaken belief that it is all very far away, on a planet visited only by graduates of the School for Space Cadets. Diagrams printed out from computers communicate a second and subtler falsehood: they lead the reader to think he is seeing the things themselves rather than pixellated approximations to them.

We have tried to solve this problem of the too far and the too near by putting our drawings in the human middle distance, where diagrams are drawn by hand. These reach out to the ideal world we can't see from the real world we do, as our imagination reaches in turn from the shaky circle perceived to the conception of circle itself.

Fuller explanations too will live in the middle distance: some in the appendix, others—the more distant excursions—(along with notes to the text) in an on-line Annex, at www.oup-usa.org/artoftheinfinite.

Gradually, then, the music of mathematics will grow more distinct. We will hear in it the endless tug between freedom and necessity as playful inventions turn into the only way things can be, and timeless laws are drafted—in a place, at a time, by a fallible fellow human. Just as in listening to music, our sense of self will widen out toward a more than personal vista, vivid and profound.

Whether we focus on the numbers we count with and their offspring or the shapes that evolve from triangles, ever richer structures will slide into view like beads on the wire of the infinite. And it is this wire, running throughout, that draws us on, until we stand at the edge of the universe and stretch out a hand.

one

Time and the Mind

Things occupy space—but how *many* of them there are (or could be) belongs to time, as we tick them off to a walking rhythm that projects ongoing numbering into the future. Yet if you take off the face of a clock you won't find time there, only human contrivance. Those numbers, circling round, make time almost palpable—as if they aroused a sixth sense attuned to its presence, since it slips by the usual five (although aromas often do call up time past). Can we get behind numbers to find what it is they measure? Can we come to grips with the numbers themselves to know what they are and where they came from? Did we discover or invent them—or do they somehow lie in a profound crevice between the world and the mind?

Humans aren't the exclusive owners of the smaller numbers, at least. A monkey named Rosencrantz counts happily up to eight. Dolphins and ferrets, parrots and pigeons can tell three from five, if asked politely. Certainly our kind delights in counting from a very early age:

> One potato, two potato, three potato, four;
> Five potato, six potato, seven potato, more!

Not that the children who play these counting-out games always get it right:

> Wunnery tooery tickery seven
> Alibi crackaby ten eleven
> Pin pan musky Dan
> Tweedle-um twoddle-um twenty-wan
> Eerie orie ourie
> You are out!

This is as fascinating as it is wild, because whatever the misconceptions about the sequence of counting numbers (alibi and crackaby may

3

be eight and nine, but you'll never get seven to come right after tickery), the words work perfectly well in counting around in a circle—and it's always the twenty-first person from the start of the count who is out, if "you" and "are" still act as numerals as they did in our childhood. We can count significantly better than rats and raccoons because we not only recognize different magnitudes but

> *know how to match up separate things with the successive numbers of a sequence:*

a little step, it seems, but one which will take us beyond the moon.

The first few counting numbers have puzzlingly many names from language to language. *Two, zwei, dva,* and *deux* is bad enough, even without invoking the "burla" of Queensland Aboriginal or the Mixtec "ùù". If you consult just English-speaking children, you also get "twa", "dicotty", "teentie", "osie", "meeny", "oarie", "ottie", and who knows how many others. Why is this playful speciation puzzling? Because it gives very local embodiments to what we think of as universal and abstract.

Not only do the names of numbers vary, but, more surprisingly, how we picture them to ourselves. Do you think of "six" as • • • • • •

or ⫶⫶ or • • • or • • • or • • ?

A friend of ours, whose art is the garden, has since childhood always imagined the numbers as lying on a zigzag path:

What happens, however, if we follow Isobel's route past 60? It continues into the blue on a straight line. Almost everyone lets the idiosyncrasies go somewhere before a hundred, as not numbers but Number recedes into the distance. 3 and 7, 11 and 30 will have distinct characters and magical properties, perhaps, for many—but is 65,537 anyone's lucky number? What makes mathematics so daunting from the very start is how its atoms accelerate away. A faceless milling crowd has elbowed out the kindly nursery figures. Its sheer extent and anonymity alienate our humanity, and carry us off (as Robert Louis Stevenson once put it) to where there is no habitable city for the mind of man.

We can reclaim mathematics for ourselves by going back to its beginnings: the number one. Different as its names may be from country to country or the associations it has for you and me, its geometric representation is unambiguous: • . The notion of one—*one* partridge, *a* pear tree, *the whole*—feels too comfortable to be anything but a sofa in the living-room of the mind.

Almost as familiar, like a tool whose handle has worn to the fit of a hand, is the action of adding. We take in "1 + 1", as a new whole needing a new name, so easily and quickly that we feel foolish in trying to define what addition is. Housman wrote:

> To think that two plus two are four
> And neither five nor three
> The heart of man has long been sore
> And long 'tis like to be.

Perhaps. But the head has long been grateful for this small blessing.

With nothing more than the number one and the notion of adding, we are on the brink of a revelation and a mystery. Rubbing those two sticks together will strike the spark of a truth no doubting can ever extinguish, and put our finite minds in actual touch with the infinite. Ask yourself how many numbers there are; past Isobel's 60, do they come to a halt at 65,537 or somewhere out there, at the end of time and space? Say they do; then there is a last number of all—call it n for short. But isn't $n + 1$ a number too, and even larger? So n can't have been the last—there *can't be* a last number.

There you are: a proof as profound, as elegant, as imperturbable as anything in mathematics. You needn't take it on faith; you need neither hope for nor fear it, but know with all the certainty of reason that the counting numbers can't end. If you are willing to put this positively and say: there *are* infinitely many counting numbers—then all those differences between the small numbers you know, and the large numbers you don't, shrink to insignificance beside this overwhelming insight into their totality.

This entente between 1 and addition also tells you something important about each point in the array that stretches, like Banquo's descendants, even to the crack of doom. Every one of these counting numbers is just a sum of 1 with itself a finite number of times: $1 + 1 + 1 + 1 + 1 = 5$, and with paper and patience enough, we could say that the same is true of 65,537.

These two truths—one about *all* the counting numbers, one about *each* of them—are very different in spirit, and taken together say something about how peculiar the art of mathematics is. The same technique of merely going on adding 1 to itself shows you, on the one hand, how each of the counting numbers is built—hence where and what each one is; on the other, it tells you a dazzling truth about their totality that overrides the variety among them. We slip from the immensely concrete to the mind-bogglingly abstract with the slightest shift in point of view.

∞

Armies of Unalterable Law

Does number measure time, or does time measure number? And in one case or both, have we just proven that ongoing time is infinite? Like those shifts from the concrete to the abstract, mathematics also alternates minute steps with gigantic leaps, and to make this one we would have to go back to what seemed no more than a mere form of words. We asked if you were willing to recast our negative result (the counting numbers never end) positively: there are infinitely many counting numbers. To put it so seems to summon up an infinite time through which they are iterated. But are we justified in taking this step?

To speak with a lawyerly caution, we showed only that *if* someone claimed there was a last number we could prove him wrong by generating—in time—a next. Were we to turn our positive expression into a spatial image we might conjure up something like a place where all the counting numbers, already generated, lived—but this is an image only, and a spatial image, for a temporal process at that. Might it not be that our proof shows rather that our imaging is always firmly anchored to present time, on whose moving margin our thought is able to make (in time) a next counting number—but with neither the right, ability, nor need to conjure up their totality all at once? The tension between these two points of view—the potentially infinite of motion and the actual infinity of totality—continues today, unresolved, opening up fresh approaches to the nature of mathematics. The uneasy status of the infinite will accompany us throughout this book as we explore, return with our trophies, and set out again.

Here is the next truth. We can see that the sizable army of counting numbers needs to be put in some sort of order if we are to deploy it. We could of course go on inventing new names and new symbols for the numbers as they spill out: why not follow one, two, three, four, five, six, seven, eight, and nine with kata, gwer, nata, kina, aruma (as the Oksapmin of Papua New Guinea do, after their first nine numerals, which begin:

tipna, tipnarip, bumrip . . .)? And surely the human mind is sufficiently fertile and memory flexible enough to avoid recycling old symbols and follow 7, 8, and 9 with @, ¤, β—dare we say and so on?

The problem isn't a lack of imagination but the need to calculate with these numbers. We might want to add 8 and 9 and not have to remember a fanciful squiggle for their sum. The great invention, some five thousand years ago, of positional notation brought the straggling line of counting numbers into squadrons and regiments and battalions. After a conveniently short run of new symbols from 1 (for us this run stops at 9), use 1 again for the next number, but put it in a new column to the left of where those first digits stood. Here we will keep track of how many tens we have. Then put a new symbol, 0, in the digits' column to show we have no units. You can follow 10 with 11, 12 and so on, meaning (to its initiates) a ten and one more, two more, . . . Continue these columns on, ever leftward, after 99 exhausts the use of two columns and 999 the use of three. Our lawyer from two paragraphs ago would remind us that those columns weren't "already there" but constructed when needed. 65,537, for example, abbreviates

ten thousands	thousands	hundreds	tens	units
6	5	5	3	7 .

As always in mathematics, great changes begin off-handedly, the way important figures in Proust often first appear in asides. Zero was only a notational convenience, but this nothing, which yet somehow is, gave a new depth to our sense of number, a new dimension—as the invention of a vanishing point suddenly deepened the pictorial plane of Renaissance art (a subject to which we shall return in Chapter Eight).

But is zero a number at all? It took centuries to free it from sweeping the hearth, a humble punctuation mark, and find that the glass slipper fitted it perfectly. For no matter how convenient a notion or notation is, you can't just declare it to be a number among numbers. The deep principle at work here—which we will encounter again and again—is that something must not only act like a number but interact companionably with other numbers in their republic, if you are to extend the franchise to it.

This was difficult in the case of zero, for it behaved badly in company. The sum of two numbers must be greater than either, but $3 + 0$ is just 3 again. Things got no better when multiplication was in the air. $3 \cdot 17$ is different from $4 \cdot 17$, yet $3 \cdot 0$ is the same as $4 \cdot 0$—in fact, anything times 0 is 0. This makes sense, of course, since no matter how many times you add nothing to itself (and multiplication is just sophisticated addition, isn't it?), you still have nothing. What do you do when someone's services are vital to your cause, for all his unconventionality? You do what the French did with Tom Paine and make him an honorary citizen. So zero joined the republic of numbers, where it has stirred up trouble ever since.

Our primary mathematical experience, individually as well as collectively, is counting—in which zero plays no part, since counting always starts with one. The counting numbers (take 17 as a random example), parthenogenetic offspring of that solitary Adam, 1, came in time to be called the *natural* numbers, with ℕ as their symbol. Think of them strolling there in that boundless garden, innocent under the trees. For all that we have now found a way to organize them by tens and hundreds, they seem at first sight as much like one another as such offspring would have to be. Yet look closer, as the Greeks once did, to see the beginnings of startling patterns among them. Are they patterns we playfully make in the ductile material of numbers, as a sculptor prods and pinches shapes from clay? Or patterns only laid bare by such probing, as Michelangelo thought of the statue which waited in the stone? Of all the arts, mathematics most puts into question the distinction between creation and discovery.

If you happened to picture "six" this way •   • , its pleasing triangu- • • • lar shape might have led you to wonder what other natural numbers

were triangular too. Add one more row of dots— • • • so 10 is • • • • triangular. Or take a row away— : 3 is triangular too. 3, 6, 10 . . . 15 would be next, by adding on a row of five dots to the triangular 10; then comes 21. We might even be tempted to push the pattern back to one, • , as if it were a triangular number by default (extending the franchise again).

Here are the first six triangular numbers:

| 1 | 3 | 6 | 10 | 15 | 21 |

Each is bigger than the previous one by its bottom row, which is the next natural number. This pattern clearly undulates endlessly on.

Idly messing about—the way so many insights burst conventional bounds—you might ask what other shapes numbers could come in: squares, for example. 4 is a square number: 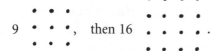 and the next would be

9 , then 16 .

Again, by courtesy, we could extend this sequence backward to 1: •. The first six square numbers, each gotten by adding a right angle of dots to the last,

are 1, 4, 9, 16, 25, 36. Another endless rhythm in this landscape.

But isn't all this messing about indeed idle? What light does it shed on the nature of things, what use could it possibly be?

Light precedes use, as Sir Francis Bacon once pointed out. Think yourself into the mind of that nameless mathematician who long ago made triangular and square patterns of dots in the sand and felt the stirrings of an artist's certainty that there must be a connection between them:

If there was, it was probably well hidden. Perhaps he recalled what the Greek philosopher Heraclitus had said: "A hidden connection is stronger than one we can see." Hidden how? Poking his holes again in the sand, looking at them from one angle and another, he suddenly saw:

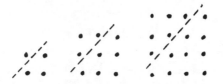

each of these square numbers was the sum of two triangular ones! Then the leap from seeing with the outer to the inner eye, which is the leap of mathematics to the infinite: this must *always* be so.

Our insight sharpens: the second square number is the sum of the first two triangular numbers; the third square of the second and third triangulars, and so on. You might feel the need now for a more graceful vessel in which to carry this insight—the need for symbols—and make up these:

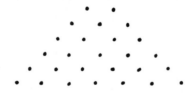

so

where that "always" is stored in the letter "n" for "any number".

By itself this is a dazzling sliver of the universal light, and its discovery a model of how mathematics happens: a faith in pattern, a taste for experiment, an easiness with delay, and a readiness to see askew. How many directions now this insight may carry you off in: toward other polygonal shapes such as pentagons and hexagons, toward solid structures of pyramids and cubes, or to new ways of dividing up the arrays.

As for utility, what if you wanted to add all the natural numbers from 1 to 7, for example, without the tedium of adding up each and every one? Well, that sum you want is a triangular number:

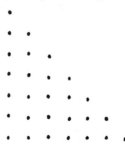

We might try writing $\triangle_7 = \boxed{7} - \triangle_6$ and work our way backward—but this will get us into an ugly tangle—and if it isn't beautiful it isn't mathematics. Faith in pattern and easiness with delay: we want to look at it somehow differently, with our discovery of page 9 tantalizingly in mind. A taste for experiment and a readiness to see askew: well, that triangle is part of a square in having a right angle at its top—what if we tilt it over and put the right angle on the ground:

Why? Just messing about again, to make the pattern look square-like; but this feels uncomfortable, incomplete—it wants to be filled out (perhaps another ingredient in the mix of doing mathematics is a twitchiness about asymmetries).

If we complete it to a square, we're back to what proved useless before. Well, what about pasting its mirror image to it, this way?

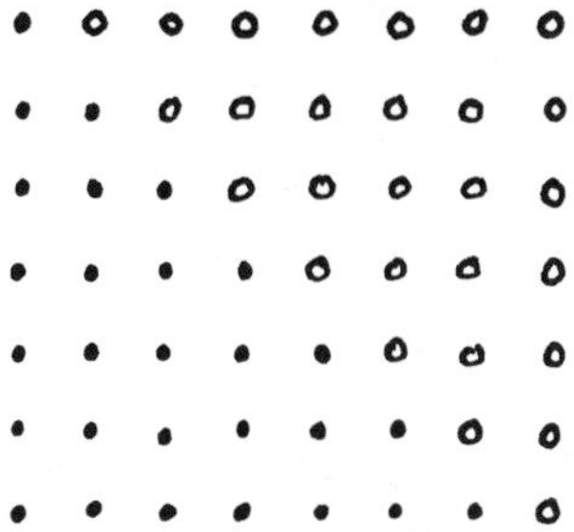

This doesn't give us a 7 · 7 square but a 7 · 8 rectangle . . . and we want only the unmirrored half of it—that is, $\frac{(7 \cdot 8)}{2}$, which is—28! Is this it?

Is $1 + 2 + 3 = \dfrac{(3 \cdot 4)}{2}$? Yes, 6.

And $1 + 2 + 3 + 4 + 5 + 6 + 7 + 8 = \dfrac{(8 \cdot 9)}{2} = 36$!

Another way of saying—or seeing—this: in order to find the sum of the first seven numbers, $\triangle$, we took $\triangle$ and slid another $\triangle$ next to it, *upside-down*—then took (of course!) half the result. When you straighten out the triangles $\triangle \, \triangledown$ you get a 7 by 8 rectangle, half the dots in which give the desired 28.

So in general,

$$1 + 2 + 3 + \ldots + n = n \cdot \frac{(n+1)}{2}$$

for any number n.

Or

$$\triangle_n = \frac{\triangle_n \cdot \triangledown^n}{2} = \frac{n \boxed{\,n+1\,}}{2} .$$

The sum of the first hundred natural numbers, for example, must be $\frac{(100 \cdot 101)}{2} = 5{,}050$. Experiments of light have yielded, as Bacon foretold, experiments of fruit.

∞

Unnatural Numbers

We could play in these pastures forever and never run out of discoveries. But something came up in passing, just now, which suggests that a neighboring pasture may have grass uncannily greener. Our abandoned attempt to find $\triangle$ by looking at $\boxed{7} - \triangle$ brought up subtraction, which isn't at home among the natural numbers.

Aren't negative numbers in fact ridiculously unnatural? Five-year-olds—fresh from the Platonic heaven—will tell you confidently that such numbers don't exist. But after a childhood of counting games, years of discretion approach with the shadows of commerce and exchange. I had three marbles, then lost two to you, and now I have one. I lose that one and am left with none, so I borrow one from a friend and proceed to lose that too, hence owing him one. How many have I? Even recognizing that I had one marble after giving up two is scaly, a snake in our garden, the presage of loss.

How are we even to picture the negative numbers—by dots that aren't there?

> Yesterday upon the stair
> I met a man who wasn't there.
> He wasn't there again today—
> I wish that man would go away.

But the negative numbers won't go away: Northerners are intimately familiar with them, thanks to thermometers, and all of us, thanks to debt.

Perhaps by their works shall you know them, through seeing the palpable effects of *subtracting*. Look again at our triumphant discovery of what the first n natural numbers added up to. If we subtract from these numbers all the evens, what sum are we left with—what is the sum of the odds?

$$1 + 3 = 4$$
$$1 + 3 + 5 = 9$$
$$1 + 3 + 5 + 7 = 16$$
$$1 + 3 + 5 + 7 + 9 = 25$$

It looks as if it might be the square of *how many odd numbers we are adding*. And here is a wonderful confirmation of this, in the same visual style as our last one—another piece of inspired invention. When we add right angles of dots to the previous ones, as we did on page 9,

$$1 \qquad 1+3 \qquad 1+3+5 \qquad 1+3+5+7.$$

what are we doing but adding up the successive odd numbers? So of course their sum is a square: the square of how many odds we have added up.

A thousand years of schoolchildren caught the scent of subtraction in problems like this, as they studied their *Introductio Arithmetica*. It had been written around A.D. 100 by a certain Nichomachus of Gerasa, in Judea. His vivid imagination conjured up some numbers as tongue-less animals with but a single eye, and others as having nine lips and three rows of teeth and a hundred arms.

Subtracting—taking away the even numbers from the naturals—has left us with the odd numbers. To people making change, subtraction turns into what you would have to add to make the whole ("98 cents and 2 makes 1, and 4 makes 5"). But it is an act also of adding a *negative quantity*: $5 and a debt of 98 cents comes to $4.02. Does the fact that we can't see the negative numbers themselves make them any less real than the naturals? The reality of the naturals, after all, is so vivid precisely *because* we can't sense them: numbers are adjectives, answering the question "how many", and we see not five but five oranges, and never actually see 65,537 of anything: large quantities are blurs whose value we take on faith. If we come to treat numbers as nouns—things in their own right—it is because of our wonderful capacity to feel at home, after a while, with the abstract. On such grounds the negatives have as much solidity as the positives, and ramble around with them, like secret sharers, in our thought.

We extend the franchise to them by calling the collection of natural numbers, their negatives and zero, the *Integers*: upright, forthright, intact. The letter $\mathbb{Z}$, from the German word for number, *Zahl*, is their symbol, and −17 a typical member of their kind. And once they are incorporated to make this larger state, we find not only our itch to symmetrize satisfied, but our sense of number's relation to time widened. If the positive natural numbers march off toward a limitless future, their negative siblings recede toward the limitless past, with 0 forever in that middle we take to be the present. It takes a real act of generosity, of course, to extend the franchise as we have, because we so strongly feel the birthright of the counting numbers. "God created the natural numbers," said the German mathematician Kronecker late in the nineteenth century, "the rest is the work of man." And certainly zero and the negatives have all the marks of human artifice: deftness, ambiguity, understatement. If you like, you can preserve the Kroneckerian feeling of the difference between positives and negatives by picturing our present awareness as the knife-edge between endless discovery ahead and equally endless invention behind.

∞

From Ratios to Rationals

You pretty much know where you are with the integers. There may be profound patterns woven in their fence-post-like procession over the horizon, but they mark out time and space, before and behind, with comforting regularity. Addition and multiplication act on them as they should—or almost: $(-6) \cdot (-4) = 24$: a negative times a negative turns out, disconcertingly, to be positive. Why this should—why this *must*— be so we will prove to your utter satisfaction in Chapter Three. Otherwise, all is for the best in this best of all possible worlds.

Exhilarated by its widened conception of number, mind looks for new lands to colonize and sees an untamed multitude at hand. For from the moment that someone wanted to trade an ox for twenty-four fine loincloths, or a chicken for 240 cowry shells, making sense of ratios became important. You want to scale up this 2 by 4 wooden beam to 6 by— what? Three of your silver shekels are worth 15 of your neighbor's tin mina: what then should he give you for five silver shekels?

The Greeks found remarkable properties of these ratios and subtle ways of demonstrating them. If an architect wondered what length bore the same relation to a length of 12 units that 4 bears to 7, a trip with his local geometer down to the beach would have him drawing a line in the sand 4 units long; and at any angle to that, another of 7 units, from the same starting-point, A:

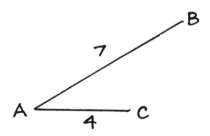

the urge for completion would lead them both to draw the third side, BC, of their nascent triangle. But now the geometer continues the lines AB and AC onward:

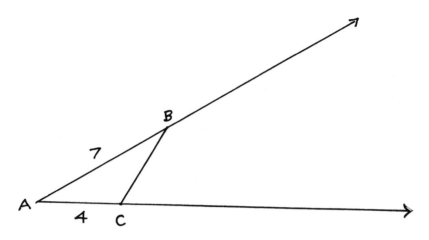

and marks a point D on AB's extension so that AD is 12 units long:

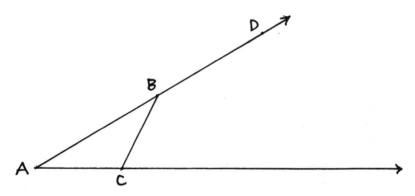

ingenuity and an intimacy with similar triangles now leads him to draw from D a line parallel to BC, meeting AC at E:

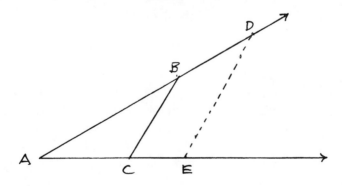

AE will be in the same ratio to 12 as 4 is to 7.

At no time, you notice, was $\frac{4}{7}$ called a number, nor was a fraction like $\frac{x}{12}$ involved; no one solved $\frac{4}{7} = \frac{x}{12}$ for x to find x equal to $\frac{48}{7}$. Those expressions couldn't be numbers to the early Greeks, for whom magnitudes were one thing, but their ratios another. Both were of vital importance to Pythagoras and his followers, who in southern Italy and Greece from the fifth century B.C. onward revealed to their initiates the deep secret that numbers are the origin of all things, and that their ratios made the harmonies of the world and its music. For if a plucked string gives middle C, then plucking a string half its length would give the octave above middle C. A string $\frac{2}{3}$ as long as the original C string would give you its fifth, G; $\frac{3}{4}$ as long, its fourth, F—those intervals that are the foundation of our scales. These ratios were propagated through the universe, making the accords that are the music of the spheres (we don't hear it because its sound is in our ears from birth). But $\frac{2}{3}$ or $\frac{3}{4}$ couldn't possibly be numbers, because numbers arose from the unit, and the unit was an indivisible whole.

How nightmarish it would have been for a Pythagorean to think of that whole fractured into fractions. It would mean that how things stood to one another—their ratios—and not the things themselves were ultimately real: and they could no more believe this than we would think that adjectives and adverbs rather than nouns were primary. That would have led to a world of flickering changes, of fading accords and passing dissonances, of qualities heaped on qualities, where shadowy intimations of what had been and what would be tunneled like vortices through a watery present you never stepped in twice.

If Greek philosophers and mathematicians did not have fractions, it seems their merchants did—picked up, perhaps, in their travels among the Egyptians, for whom fractions (though only with 1 in their numerators) dwelt under the hawklike eye of Horus.

Against this background of daily practice, insights into how ratios behaved kept growing, until inevitably they too became embodied in numbers. How could properties accumulate without our concluding that what has them must be a thing—especially since we are zealous to make objects out of whatever we experience? So they came to live among the rest as pets do among us, each with its cargo of domestic insects:

> Great fleas have little fleas
> Upon their backs to bite 'em,
> And little fleas have lesser fleas,
> And so *ad infinitum.*

For an uncanny property of these fractions is that they crowd endlessly into every smallest corner. Between any two you will always find another: $\frac{5}{12}$ lies between $\frac{1}{3}$ and $\frac{1}{2}$; between $\frac{5}{12}$ and $\frac{1}{2}$ is $\frac{11}{24}$; between $\frac{11}{24}$ and $\frac{1}{2}$ is $\frac{23}{48}$. The average of the two ends falls between them, along with how many other splinters of the whole, so that infinity not only glimmers at the extremities of thought but is here in our very midst, an infinity of fractions in each least cleft of the number-line.

So the franchise was hesitantly extended to ratios in the guise of fractions, although uneasiness at splitting the atomic unit remained. The fractions, preserving traces of their origin in their official name of *Rational Numbers*, were symbolized by the letter $\mathbb{Q}$, for "quotient". Does this variety of names reflect the doubts about their legitimacy? To counteract these worries, notice that the integers now can be thought of as rationals too: each—like 17—is a fraction with denominator 1: $\frac{17}{1}$ (or, if you have a taste for the baroque, $\frac{34}{2}$, $\frac{51}{3}$, and so on). And notice how this new flood of intermediate numbers makes number itself suddenly much more time-like: flowing with never a break, it seems, invisibly past or through us.

We can conclude: numbers are rational, and a rational is an expression of the form $\frac{a}{b}$, where a and b can be any integer. Or almost any: a pinprick of the old discomfort remains in the fact that b, the denominator, cannot be 0. Tom Paine again, waving his *Common Sense*. Why it makes sense (not so common, perhaps) that you cannot divide by zero will be part of the harvest reaped in Chapter Three.

$$\infty$$

Nameless Dread

Fractions keep crowding whatever space you imagine between them, a claustrophobe's nightmare. Thought of as ratios, however, they are a Pythagorean's dearest dream: any two magnitudes, anywhere in the universe, would stand to one another as a ratio of two natural numbers. Take the module of the way we count, the number 10. Is it a coincidence that it is the triangular sum of the first four counting numbers?

The Pythagoreans didn't think so: 10 must have seemed to them as compact of meaning as the genetic code, coiled within a cell, seems to us. For not only did the individual numbers of this triangular ten—which they called the tetractys—each carry a distinct significance (unity, duality, the triangular, the square . . .), but their ratios, as you saw, expressed the harmony which orders the universe. No wonder a Pythagorean's most sacred oath was by this tetractys, the Principle of Health and "fount and root of ever-flowing nature."

In this atmosphere their wonderful works of geometry grew: insights others may previously have had, but based now for the first time on proof. It was no longer a matter of faith. No mystical revelation, no authority human or divine, authenticated these truths. Mind confronted them directly through impartial logic, which lifted you up from the streets of Tarentum or a hill overlooking the Hellespont to the timeless topography of ideas: not the setting, you would have imagined, for the destruction of the Pythagorean attunements. Yet the tragic irony that runs beneath all Greek thought burst out most catastrophically here, for the wedding of insight to proof in Pythagoras's prized theorem—that the square on a right triangle's hypotenuse equals in area that of the sum of squares on the two sides

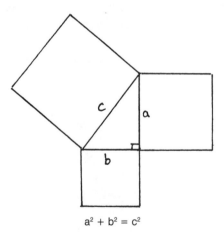

$$a^2 + b^2 = c^2$$

had a lame patricide as its offspring.

We have only the faintest echoes of the story, in late and unreliable sources at that, since secrecy obsessed the Pythagoreans generally, but at this moment most of all. A Pythagorean named Hippasus, they say, from Metapontum, used that great theorem to prove there was a magnitude which, when compared to the unit length, couldn't make a ratio of two natural numbers. But if this were so, where would the music of the spheres and the harmony of things be? Where the whole, the tetractys, the moral foundations of life?

Yet Hippasus's proof had an iron certainty to it. Put in modern terms, a right triangle both of whose legs are of length one has a hypotenuse of length h, which the theorem lets us calculate.

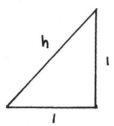

We must have $h^2 = 1^2 + 1^2$, that is, $h^2 = 2$. So the length of $h = \sqrt{2}$. You need only look at a diagram to convince yourself that h as much deserves to be called a length as do the other two sides. If it isn't a natural number, it must, for a Pythagorean, be a ratio of natural numbers a and b:

$$\sqrt{2} = \frac{a}{b}.$$

This is already a little awkward, since ratios weren't magnitudes for them, as a length would have to be. But much worse lay ahead. If this ratio wasn't in lowest terms—if a and b, that is, have some common factor like 2—cancel it out until the equivalent ratio *is* in lowest terms. Let's still denote it by a and b, knowing now that these two natural numbers have no factor in common.

Hippasus let the desire to simplify, and a craftsman's feel for arithmetic, now take him where they would. It is this artistic motivation and reckless commitment to whatever consequences follow that is the mathematician's real tetractys, the sign to kindred spirits across millennia; and it is what makes for the glories and despairs of mathematics.

" $\sqrt{2}$ " is clumsy both as a symbol and a thought. If we square both sides of our equation we come up with the simpler translation:

$$2 = \frac{a^2}{b^2}.$$

And since multiplication is in turn simpler than division, translate again by multiplying both sides of this equation by b^2:

$$2b^2 = a^2.$$

At this point the Hippasus in each of us pauses to assess what has happened. Since b is a natural number, so is b^2; and twice a natural number, such as $2b^2$, is an even number. The even and odd, like left and right, darkness and light, bad and good, were pairings immensely congenial to the Pythagoreans, so the evenness of a^2 would have struck them.

Only numbers that are themselves even can have even squares: an odd squared (such as 5) will stay odd (25). Hence since a^2 is even, a must be too, which means it is twice some natural number n:

$$a = 2n.$$

Hence $a^2 = 4n^2$.

Carry this consequence carefully back to our last equation, $2b^2 = a^2$, and replace a^2 there by $4n^2$:

$$2b^2 = 4n^2.$$

Why do this? You might think of the initial impulse as experimental; or perhaps intuition is flowing as surely as a river to the sea.

Once again let the aesthetic impulse to simplify lead our efforts, and divide both sides of this latest equation by 2:

$$b^2 = 2n^2.$$

The same reasoning as before shows us that b^2 is even—hence, so is b.

The dénouement of this drama is on us before we have time to draw breath. We have seen that a and b must both be even, so they have 2 as a common factor. But we canceled out all common factors when we began! So $\frac{a}{b}$ is a fraction which must be simultaneously in lowest terms and not in lowest terms. We followed a path and it brought us to the impossible, a contradiction—yet each of our steps was wholly logical. The only possibility left must be that assuming in the first place that $\sqrt{2}$ was rational ($\sqrt{2} = \frac{a}{b}$) was mistaken: $\sqrt{2}$ is not a ratio of natural numbers. It wasn't, isn't, and never can be a rational number; yet it clearly exists, stretched out on the hypotenuse, just as much as do unit lengths.

The Pythagoreans couldn't deny the validity of Hippasus's proof. One story has it that they were at sea when he told it to them, and they—or the gods—drowned him for his impiety. The proof they could no more drown than the infant Oedipus could be killed by his parents. Henceforth they had to live with $\sqrt{2}$ being *irrational*, or as they called it ἄλογος, *nameless*. And they lived with it in dread, like priests who perform their office knowing that God is dead. It was the secret deep within the nested Pythagorean secrets.

There it grew, for any natural multiple of $\sqrt{2}$ must be irrational also. If $7\sqrt{2}$, for example, were rational—if $7\sqrt{2} = \frac{p}{q}$—then $\sqrt{2}$ would equal $\frac{p}{7q}$, a rational again. Hippasus from Hades calls out that this cannot be. The growth metastasized: any rational whatever (except, of course, 0) times $\sqrt{2}$ will be irrational, since if $(\frac{a}{b}) \cdot \sqrt{2} = \frac{p}{q}$, then $\sqrt{2} = \frac{bp}{aq}$, which is a ratio of natural numbers. The tight line of the rationals was now peppered with these irrational offspring of $\sqrt{2}$.

The darkness grew only deeper. $\sqrt{3}$ turned out to be irrational also (the proof is very like Hippasus's, but with a threefold classification of naturals instead of the twofold distinction we had for $\sqrt{2}$). So, therefore, were all its numerous progeny. Then $\sqrt{5}$ followed suit, and $\sqrt{6}$, and $\sqrt{7}$. In fact if a natural number wasn't a square like those we saw on page 9, its square root had to be irrational. Swarms of irrationals were now loose in the land, with plagues to follow: cube roots of numbers not perfect cubes are irrational too, and fourth roots of numbers not perfect fourths (the fourth root, for example, of 81, $\sqrt[4]{81}$, is 3, but $\sqrt[4]{80}$ and $\sqrt[4]{82}$ are irrational)—and so terrifyingly on.

The terror lies in what seems our inability to accommodate all these invaders. Remember how packed the line of rational numbers was to begin with—as densely settled as the fabled midwestern town whose built-up zones had a house between any pair of houses. The rationals are dense, as we saw before, with a rational (their average, for example) between any two rationals. Where then could all those irrationals possibly fit?

If you claim they aren't on the number line at all, gently lower the hypotenuse of the triangle we began with, as if it were the boom of a crane, until it rests on the line:

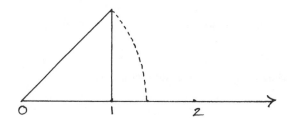

Its tip touches the line at a point somewhere between 1 and 2 (between 1.4 and 1.5 if you care to be more exact, or even more precisely, between 1.41 and 1.42), so this point has the irrational number $\sqrt{2}$ as its address.

We will never be at ease with this, but at least we can try to grasp the situation in another way: through decimals. If you turn a rational number into a decimal, that decimal will either peter out eventually to nothing but zeroes ($\frac{1}{2} = 0.5000\ldots$ —or we could put a bar above the 0 to show it repeats forever: $0.50\overline{0}$) or it will begin to repeat. So $\frac{1}{3} = 0.333\ldots$ that is, $0.\overline{3}$, and $\frac{1}{7} = 0.142857142857142857\ldots$ that is, $0.\overline{142857}$. *Why must this be?* Because you get the decimal by dividing the denominator into the numerator, and at each step you get a remainder. If you are dividing by seven, the only possible remainders are 0, 1, 2, 3, 4, 5, and 6 (if you get a larger remainder, you could have divided 7 in one more time). How many different remainders are there? Seven: there can't be any more. This means that after a while the remainders start recycling:

$$\frac{1}{7} = 7\overline{)1.0000000000}\ \ldots \quad 0.1428571$$

$$
\begin{array}{r}
0.1428571 \\
7\overline{)1.0000000000}\ \ldots \\
\underline{-7} \\
③0 \\
\underline{-28} \\
②0 \\
\underline{-14} \\
⑥0 \\
\underline{-56} \\
④0 \\
\underline{-35} \\
⑤0 \\
\underline{-49} \\
①0 \\
\underline{-7} \\
③
\end{array}
$$

and we see the cycle beginning again.

Clearly the very nature of division forces the decimal representation of a fraction to repeat. So if a decimal *doesn't* repeat, it can't represent a rational number!

This tells us two very striking things at the same time. First, that because $\sqrt{2}$ is irrational, its decimal form can't repeat:

$$\sqrt{2} = 1.41421356\ldots$$

No matter how long we go on, no cycle will emerge. Hippasus's proof that $\sqrt{2}$ is irrational guarantees that $\sqrt{2}$ lies not just between 1.41 and 1.42, but between 1.414 and 1.415, between 1.4142 and 1.4143, and so on. Squeeze it as tightly as you like between two rationals, it will squeak and scurry away down an infinite sequence of ever-narrowing cracks.

The second thing it tells us is that we needn't confine ourselves to various roots to find irrationals: we can now produce them at will. All we need do is manufacture a decimal that never repeats. How do that in a finite lifetime? We can't just start writing out arbitrary strings of digits after a decimal point:

$$0.180094051\ldots$$

because we will eventually stop, and nothing guarantees that our string won't now or at some future time begin to replicate. As a matter of fact, the string above is the beginning of the decimal representation of a perfectly good rational number,

$$\frac{429376501}{2364179256}$$

which, because of the size of its denominator, needn't start repeating for more than two billion decimal places.

We need a guarantee *in the way we make it* that our decimal can't repeat. In the midst of the chaos mind has released, the power of mind to make order at one remove—its power over the infinite—emerges too. For we can build into the very instructions that will produce our decimal, the guarantee that it cannot repeat, so that it is indeed an irrational number.

Picture a computer that will print out digits forever, one by one, after an original 0 and decimal point. We program it with only three instructions:

1. print 5.
2. print one more 6 in a row than were printed before the "5" of the previous step.
3. return to step 1.

Once we set the machine in motion it prints "5" after the 0. initially there, giving us

    0.5

then, because there were *no* 6s printed before the 5, it prints one 6:

    0.56

and cycling back to its first order, prints 5 again:

    0.565

now it will print two 6s

    0.56566

then a 5, then three 6s

    0.565665666 . . .

You see the pattern of this non-repeating pattern: the strings of sixes grow ever longer, and no cycle can possibly occur. We have, with a few words, cast an infinite line with an irrational hooked on its receding end—an irrational which has a unique location, somewhere between $\frac{56}{100}$ and $\frac{57}{100}$.

The irrationals that such an algorithm can generate are mind-bogglingly infinite in number: we could use any digits other than 5 and

6; we could alter the instructions for the lengths of successive strings; we could put any integer we like before the decimal point. The rationals are everywhere—the irrationals are everywhere else.

Taken all together, the rationals and the irrationals came to be called the *Real Numbers*, denoted by $\mathbb{R}$. Extending the franchise to them all means that from a distant enough standpoint they look alike: any one of them can be expressed as a decimal (17, after all, is shorthand for $17.\bar{0}$); some end in zeroes, some repeat, others are wild. They act and react with each other according to the old rules of combination, which means adding and subtracting, multiplying and dividing, and taking roots. But why call them *real*? Are they as real as this page or the light falling on it—or perhaps even more real, outlasting all? We come and go, but $\sqrt{2}$ and its ilk remain forever, and past them, the deep principles that show in their constellations. Perhaps we call them real because only now does their ensemble fully imitate time and space in their apparent continuity, or because we sense that reality ever escapes our rational convergings.

$$\infty$$

Mind and the Imagination

You may find yourself now in the distracted state where mathematicians notoriously live. The genie you rubbed from its bottle was much more powerful than you thought: barely under control. You see not only its huge, escaping shape, but—through the swirls—portents of forms even more inimical. And yet you do have a sort of authority over these numbers, since you can call irrationals from the vasty deep by such algorithms as you just saw. It is like being on how's-it-going terms with the local mob. The mathematician John von Neumann once said that in mathematics we never understand things but just get used to them. That can't be quite right—yet our understanding must be stretched to the breaking point before it becomes flexible enough to adjust to the unthinkable.

First you begin to doubt the reality of the reals. Are they actually already out there, each in its infinite splendor? Or have we instead only a machine that can mint them on demand, but with their edges shaved to varying tolerances? Thought of so, the mind resembles a totalitarian state, owning the means of production but with unregenerate individualism corrupting its inventories.

One glance, however, at the stroke of a line across a piece of paper reminds you that there—or if not there, then in what that line stands for—all of the points fully exist, rationals and irrationals alike. How could we calculate any length were our ruler not brought right up against what is, taking its measure? Even if our measurements require astronomical instruments, the distance from here to Alpha Centauri hasn't waited for us to bring it into being. The irrationals lay undiscovered in the body of mathematics as the system of tectonic plates lay undiscovered in the earth's until recently: both were there to be found, and who knows what other systems may still operate unknown?

Then you think to yourself: with just a handful of digits—some before a decimal point and some after—I can invent a number most likely never thought of before. Invent or discover, discover or invent—or do numbers evolve organically, like forms of life, when demands and conditions coincide?

Since it was those decimally advantaged numbers, the irrationals, that provoked these thoughts at the edge of reason, the weight of our perplexity falls on them. Should we really have accepted their existence with such docility? All Hippasus showed was that $\sqrt{2}$ *wasn't rational*—why grant that it *is* something else, something at all, and not just a minute gap in the number line? When we lowered the $\sqrt{2}$ hypotenuse a few pages back, perhaps its tip hovered over a hole. Why might the number line not turn out, on sufficient magnification, to be porous?

Let us focus the lens instead on how we have come up with our numbers. "One" seems there in the mind and its world, from the very start, and zero as well: something and nothing. The action of adding then gives us the naturals. Subtracting brings the negatives into the light; dividing, the rationals: and it is only when a new operation appears—the taking of roots—that the irrationals show themselves. So there we are: new numbers devised—or revealed—by operations on old ones; the familiar actions with their touring company of actors, a complete set of plots and all the dramatis personae needed to enact them.

Or *is* it complete? Can all of our cast really perform in all of the scenes? What about taking roots of negative numbers, such as $\sqrt{-1}$? This symbol stands for a number which times itself is negative one. Such a number can't be positive, since a positive times a positive is positive. It can't, however, be negative, because as you remember and Chapter Three will attest, a product of two negatives is positive too (going on with the story, while putting a proof of one of its claims on hold, is part of that easiness with delay we spoke of earlier).

Nowhere, then, on our real line—not at zero, nor to its left nor to its right, not sheltered among the rationals, nor masquerading as an irrational—can there be any number which is the square root of negative one. It is at this point that a deep quality of the mathematical art emerges—let's call it the Alcibiades Humor. For Alcibiades was the *enfant terrible* of ancient Athenian life at the time of Socrates: handsome and willful, outrageous and heroic, arrogant and playful, disrupter of discourse and envoy of passion to the feast of reason. Plutarch tells us that even as a boy, dicing in the street, he dared an angry carter to run him over—and of course the carter turned back.

The Alcibiades Humor in mathematics is just this hubris, this refusal to stop playing when all seems lost. No square root of negative one? Then let's make it up! For imagination extends beyond the real. Give this new number a name and its habitation will follow. Call it i, for imaginary; let it be a number, a new sort of number whose only property is that its square is −1:

$$ i \cdot i = i^2 = (\sqrt{-1})^2 = -1. $$

Now tightrope thinking begins, that odd blend of eliciting and inventing at the heart of mathematics, extending the frontier and the franchise. With so little to go on, what can we ask? In the spirit of i^2, see what i^4 would have to be:

$$ i^4 = i^2 \cdot i^2 = (-1) \cdot (-1) = 1 , $$

so i is, astonishingly, a fourth root of 1!

And i^3?

$$ i^3 = i^2 \cdot i = (-1) \cdot i = -i . $$

Now we have a little table of powers of this what-you-will:

$$ i^1 = i $$
$$ i^2 = -1 $$
$$ i^3 = -i $$
$$ i^4 = 1 $$

and i^5? That is $i^4 \cdot i$, or $1 \cdot i$, so i again—and now we know that the pattern of our table will cycle forever, allowing us to calculate what any power of i must be. Just divide the power you have in mind by 4 (thus casting out the cycles) and see what remainder is left—how powerful, in mathematics too, the saved remnant often is. Take i^{274}, for example; 4 into 274 leaves a remainder of 2, so i^{274} is the same as i^2, or −1.

We bring this alien slowly to earth by asking it to engage with the terrestrials. i + i is 2i, and 13i means i added to itself 13 times. 13 + i is just ... 13 + i : the alien mixes with the natives on formal terms, keeping his distance. In that remoteness he generates further imaginaries, as 1 generated the natural numbers. On a trajectory of their own they range and play, as addition, subtraction, multiplication, and division draw them endlessly out:

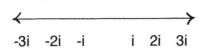

But just in the midst of these eccentric, playful creatures is 0i, and that is 0: a *real* number! It is where the trajectory of i strikes the real line, so that we needn't picture these two progressions as parallel or skew, but intersecting—and therefore (so much created out of nothing and imagination) producing a plane of numbers where once a thin line had been. The real line and the imaginary line need not, of course, meet at right angles, but to give some familiarity to the representation of space, it's convenient to work with the one unique angle that divides the universe into four equal quadrants (and while we are at it, to keep the grid square by letting the units on both lines be the same length).

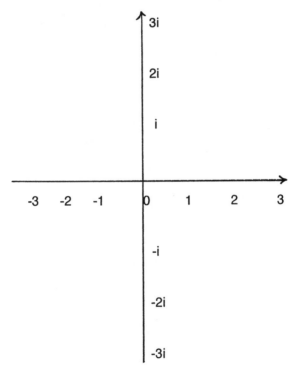

The Complex Plane

It might be a mistake to pause now and ask what these imaginaries *really* are. They had been described as "sophistic" by Italian mathematicians in the Renaissance; it was Descartes who dismissively first called them "imaginary". Newton held them to be impossible, and Leibniz said that $\sqrt{-1}$ was an amphibian between being and not-being. In 1629 the Frenchman Albert Girard agreed: "Of what use are these impossible solutions? I answer: for three things—for the certitude of the general rules, for their utility, and because there are no other solutions to certain equations." Their use—how they behave, what they tell us about numbers and the mind and the world—will be our way to understand them better, for use and understanding combine in complex solutions to questions we ask too simply.

In fact, the combinations of real and imaginary numbers—hybrids such as $3 + 5i$ or $\sqrt{7} - 4i$—are called *Complex Numbers* with the letter $\mathbb{C}$ on their caps. 17, for example, is shorthand for the complex number $17 + 0i$.

Do you feel we have been hustling you through inadequate justifications, like confidence-tricksters more eager to persuade than explain? We can't after all just say that anything we choose is a number, or argue like lawyers from precedent, or like prophets, from revelation. We have to show that the franchise has been legitimately extended to these imaginaries, and that they can do work that none of the other citizens could manage. But this we must do in the context of mathematical legitimacy itself, which is the subject of the next chapter.

Like Sheherazade, let's end one story as the next begins to edge forward and say that our operations and what they operate on are at last complete: the natural numbers are nested inside the integers, those in the rationals and those within the reals—and the reals are no more than a line on the infinite complex plane we drew on page 27. Were there Pythagoreans today, these nests might serve instead of the tetractys as the fount and root of ever-flowing nature:

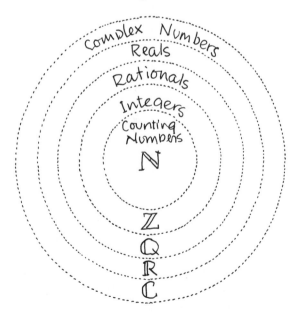

The Talisman of the New Pythagoreans

two

How Do We Hold These Truths?

We knew as soon as we saw it that the sum of the first seven natural numbers was half the number of dots in a 7 by 8 rectangle—and also saw at once that this must be true in general:

$$1+2+3+ \ldots +n = \frac{n \cdot (n+1)}{2}.$$

Of course this was a sophisticated kind of seeing, done less with the body's eye than the mind's, focused at infinity. Somehow, inexplicably, we seem to jump in a moment out of time—or is it into a sort of time with the breadth of space—is that where these figures lie?

We could show off by applying our insight to particular numbers:

$$1+2+3+ \ldots +81 = \frac{(81 \cdot 82)}{2} = 3321.$$

Such examples might strengthen our conviction, but no matter how many of them encourage our belief, there are too many numbers for examples ever to prove anything. The claim that no natural number is greater than a million is, after all, confirmed by the first million test cases.

It is easy to *ask* how we know that a statement is always true, but very hard to answer. A twelfth-century Indian proof of the Pythagorean Theorem consists of no more than two puzzle-like diagrams with the single explanatory word: "Look!" And on the next page is a thoroughly wordless early proof from China.*

---

*These diagrams given in silence have the air of a rite of passage. The initiate must first remark that since the two squares are the same size, their areas are equal; then, that the four triangles in each, although differently situated, are all identical—hence the area remaining in each square after removing them is the same. But that area is made up in the first diagram of two squares, one on each of the triangle's legs; and in the second, of a square on its hypotenuse—so that the sum of the areas of the two squares equals the area of the third.

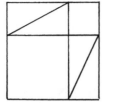

And here indeed, looking leads to seeing. Is this because an exemplar rather than an example—a particular case whose particularity doesn't matter—wakens our sense of analogy and the ability to recognize pattern?

In order to savor once more this all too fugitive experience, here is a very different way of seeing that

$$1+2+3+ \ldots +n = \frac{n \cdot (n+1)}{2}.$$

Again we choose an example—say, 10. You look at the sum

$$1 + 2 + 3 + 4 + 5 + 6 + 7 + 8 + 9 + 10$$

and ingrained habits of reading from left to right, as well as being systematic, lead you to starting: 1 plus 2 is 3, and 3 makes 6, and 4 makes 10. . . But what if you look at it differently (and the secret of all mathematical invention is looking from an unusual angle)—what if you add in pairs as follows:

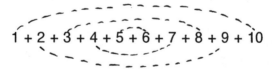

$1 + 10 = 11, 2 + 9 = 11, 3 + 8 = 11$—in fact, all these pairs will add up to 11! And how many pairs are there? 5—that is, half of 10. So

$$1+2+3+4+5+6+7+8+9+10 = \left(\frac{10}{2}\right) \cdot (11),$$

$$\text{or } \left(\frac{n}{2}\right) \cdot (n+1).$$

Some people relish the geometric approach, some the symbolic. This tells you at once that personality plays as central a role in mathematics as in any of the arts. Proofs—those minimalist structures that end up on display in glass cases—come from people mulling things over in strikingly different ways, with different leapings and lingerings. But is it always from the same premises that we explore? Is there some sort of *common* sense that is to reason what Jung's collective unconscious used to be to the psyche? One of these approaches, or some third, must have been in the mind of the ten-year-old Gauss—the Mozart of mathematics—when, in his first arithmetic class, he so startled his teacher. It was 1787 and Herr Büttner was in the habit of handing out brutally long sums like these, which the children had to labor over. When each one finished he added his slate to the pile growing on the teacher's desk. The morning might well be over before all had finished. But Gauss no sooner heard the problem than he wrote a single number on his slate and banged it down. "Ligget se'!" he said, in his Braunschweig accent: "There it lies!" And there it lay, the only correct answer in the lot.

Gauss may have had better access to his intuition than most of us do, but isn't it clear that what is common to us all is this very intuition? Yet ever since the earth turned out not to be flat, our trust in the obvious has been weakened. Insight and intuition were knocked off their pins by Hippasus: his irrational shook the Greeks more profoundly than the eruption of irrational passions through the sunlit surface of life.

Fear begets law. The jurist in the soul demands system to hem in the disorder that the irrationals let loose. The remedy that Eudoxus, one of Plato's followers, came up with in the fourth century B.C. was to build up even the most banal certainty on an armature of proof. This meant deducing results by pure logic from as trim and tight a foundation as he could find. These foundations were "axioms", like the familiar "equals added to equals make equals"—statements so weighty and worthy of belief that we don't even know how to doubt them. Their evolution is curious, because we are such inveterate doubters.

Plato's theory of recollection explained why we simply recognize truths for what they are: the soul had seen them directly in its abstracter state, among the eternal Ideas, before we were born. Aristotle hedged these bets: some first principles were common to all the sciences, some were justified by the consequences they begot. All came from generalizing what we saw in the physical world. The Stoic philosophers a century later spoke of a "recognizable impression" which gave us our basic certainties. Our apprehension first encounters an image as an open hand would an object; then begins to close around it in assent; next grasps it tightly—the fit of hand to object was "recognition"—and finally (here the Stoic Zeno, teaching his students, would cap his clenched right fist with his left hand) holds it as knowledge.

When the Gnostics fastened onto the Pythagorean pairing of darkness and light, putting it at the heart of everything, a belief began to grow in something on a different plane from our animal instincts: an inner or natural light which enhaloed the truth. By the time of St. Augustine and later St. Thomas Aquinas, the two strands of recognizable impression and natural light twined around each other to redefine "intuition", which gave us immediate truth.

Immediate: that was the test; and where but in France, so charmed by *élan*, would immediacy be an irresistible force? Where but in France would the graceful sweep of articulate thought guarantee its validity? On November 10, 1619, the young Descartes had a dream in the midst of the Lowland Wars, where he served with Prince Maurice of Orange. In it he saw that authority counted for nothing in mathematics, whose methods were able to find unimpeachable truths. When he wrote up the principles of this method nine years later, in *Règles pour la direction de l'esprit*, he said that in order to gain knowledge we must begin with what we can intuit clearly and immediately, pass one by one through all the relevant stages in a continuous and uninterrupted movement of thought, to see in the end the truth directly and transparently.

René Descartes (1596–1650), whose interest in mathematics was sparked by a problem he saw posted on a wall in Holland in 1618.

This trumpet call echoed as resonantly through France as playing up and playing the game did through England. You hear it in 1810, when the French geometer Gergonne wrote that axioms were theorems whose mere statement sufficed for recognizing their truth. You hear it at the end of the nineteenth century in Rimbaud's *Une saison en enfer*: "We are dedicated to the discovery of divine light. All the filthy memories are disappearing. . . . I will be allowed to possess the truth in a single soul and body." (There is also an echo here of the Stoic and Cartesian concern for purity, lest even intuition fall into error.)

But what if metaphors of light or appeals to something as flighty as imagination struck you as too flimsy a framework for the tower of mathematics? A different prospect, from the world rather than the mind, opened up with Locke and the enlightenment: a prospect whose vanishing-point was *self-evident truth*. A paradox incriminates itself without any help from others. Doesn't a tautology, at the other extreme, exonerate itself? You could doubt that the stars have fire, but there is no way of doubting that a star is a star. Such truths, which literally "say themselves", were seized on as axioms with Jeffersonian vigor.

Then came Kant. With one brilliant stroke he cut to the heart of the matter, the fundamental peculiarity of mathematics: whatever we invent in it at once seems independent of our inventing, as if it had instead been discovered to have been there even before experience. This, he said, was because we got our experiences from knitting our perceptions together in causal fabrics—but those perceptions had first been shaped by our intuitions of space and time. It was this shaping that mathematics studied (space in geometry, time in number), so of course these forms registered as *a priori:* prior to experience. They were the one and only way that mind made its perceptions (as the Stoic hand fitted its grasp to what it encountered). The basic truths—the axioms of mathematics—must therefore generate the unique set of conclusions that follow in our probing of how these intuitions of space and time work.

∞

The Tablets of the Law

What were these truths—common, self-evident, *a priori,* simple, or immediate—that mind apprehended in its out-of-mind state? They were aphorisms such as this: the order in which you add or multiply two numbers makes no difference; the result will always be the same:

$$a + b = b + a$$
$$a \cdot b = b \cdot a.$$

Only someone bamboozled by the old shell game could doubt these *Commutative Laws,* which you see in action whenever you watch the wheeling formations of a marching band. 7 columns 4 abreast turn at the drum major's whistle into 4 columns of 7. Since these laws hold for the natural numbers, the impulse of the time was to carry them through to the outermost circles of the mathematical empire, past integers and rationals, as satraps once carried the laws made in Persepolis to every Persian province.

You find these axioms stated with growing sophistication during the eighteenth and nineteenth centuries. In Germany, while Georg Ohm in the 1820s was drafting his law that united electrical voltage, current, and resistance, his younger brother Martin was making the laws for weaving the numbers together through the operations on them, such as the *Associative Law*, which declared that regrouping couldn't change a sum or a product:

$$a + (b + c) = (a + b) + c$$

$$a \cdot (b \cdot c) = (a \cdot b) \cdot c \, .$$

It was all very well and wonderfully concise to express these laws about numbers with letters, but how could we guarantee in a republic rather than a monarchy that the letters could stand for any kind of number at all? In England a man named George Peacock, who seemed able to believe six impossible things before breakfast, stated Peacock's Principle of Permanence in 1833: "Whatever form is algebraically equivalent to another form expressed in general symbols, must continue to be equivalent, whatever those symbols denote." So if an operation made sense for the natural numbers, it must—by Peacock's Principle—make sense for any kind of number. His Principle never stooped to ask why this should be so, and in fact (as we shall see on page 93), led to nonsense. Hidden in the neutral word "form", however, was the embryo of an abstractly formalist point of view that would utterly shift our understanding of mathematics.

George Peacock (1791–1858)

This changing way of looking was part of the broader Romantic rebellion against Enlightenment ideals. On his twenty-first birthday, June 8, 1831, Robert Schumann wrote in his diary: "It sometimes seems to me that my objective self wanted to separate itself completely from my subjective self, or as if I stood between my appearance and my actual being, between form and shadow." Form allied to appearance, actual being to shadow: a disturbing pairing that catches not only the split in Schumann's personality and in his ghost-ridden music, but in the time itself. It became possible to think that mathematics might rid itself of the subjective, of intuition, and find its justifications in form: in appearances that had nothing more to them than their representing this form and displaying its impersonal, formal rules. An extra incentive came from the growing fascination with ingenious artifice, with clockwork that could imitate or even surpass the organic (the mood had been set two centuries before when Pascal invented the first "mathematical engine" and remarked in awe that although it showed no trace of will, as animals did, it approached nearer to thought than all the actions of animals). Even the American Constitution, although kept folded up in a little tin box, was fondly thought of as a machine that would go of itself.

So Peacock's Principle of Permanence extended, on such alluringly formal grounds, the Commutative and Associative Laws to all kinds of numbers, as well as extending an important axiom that, in Ohm's style, tied together the operations of addition and multiplication:

$$a \cdot (b + c) = a \cdot b + a \cdot c.$$

This *Distributive Law* says that you will get the same result if you first add two numbers (b and c) and then multiply them by a third (a), or first multiply each by a and then add the results.

Formalism—where relations hold among symbols that need have no further referents—became an ideal shelter in the revolution that was sweeping through mathematics itself in the nineteenth century. Everyone had taken for granted, over the past two millennia, that Euclid's geometry described this precious only endless world in which we say we live—or in Kant's terms, the way mind *must* spatially conceive. Yet now geometries were being invented by Frenchmen and Russians, Hungarians and Germans, each different from Euclid's but as cogent. Where had the uniqueness gone? Wouldn't the fall of Geometry's house bring Arithmetic's down with it? One attempt followed another to end the scandal and purify Euclidean geometry of its vulnerable elements; but as in the French revolution that preceded this one, the cry kept sounding for an ever purer to purify the pure.

If Formalism couldn't save shape, it would save number. The axioms began to coalesce, going on from the Associative, Commutative, and Distributive—now elevated to the hollowly dignified status of "Laws"—to include something important which had been omitted until then. For those laws had said that *if* you had such-and-such numbers, then such-and-such configurations held among them. But what guaranteed that there were any numbers at all? An axiom was needed to assert that there *was* something—and why not another axiom, while we were at it, about nothing as well? Axioms, that is, which stated that "1" existed, and affirmed also the existence of "0".

Questions about how to consider mathematical existence became the special concern of David Hilbert, a German mathematician whose outlook dominated much of the twentieth century. His was an existence haunted by existence. At a meeting of mathematicians in Leipzig shortly after World War I, Hilbert asked a young Hungarian whether one of his colleagues was still alive. Yes, the Hungarian answered, and began to say where he was teaching and what he was working on. Hilbert kept interrupting: "But—"; the Hungarian went on: "And he was married a few years ago, and has three children, the oldest—" Hilbert burst out: "But I don't want to know all of that! I just asked: does he still exist?"

On the Portmanteau, if not the Permanence, Principle, 1 and 0 were packed with other significant properties as well; and despite zero's late entry into history, by the early twentieth century 1 and 0 were installed together at the beginning of mathematical creation, like Adam and Eve.

> *The axiom of additive identity:* There is a number, "0", which when added to any other leaves the sum unchanged:
>
> $$a + 0 = a \,.$$
>
> *The axiom of multiplicative identity:* There is a number, "1", which when multiplied by any other leaves the product unchanged:
>
> $$a \cdot 1 = a \,.$$

And since it isn't self-evident that these germinal numbers are different, we have to legislate it in by adding to the last axiom:

$$\text{and } 1 \neq 0 \,.$$

Each of these two axioms calls up a sibling that assures us we can come back to 0, or 1, from just about anywhere on the number line.

> *The axiom of additive inverses:* For any number a, there is another number, written –a, such that
>
> $$a + (-a) = 0 \,.$$
>
> *The axiom of multiplicative inverses:* For any number a, except 0, there is another number, written $\frac{1}{a}$, such that
>
> $$a \cdot \frac{1}{a} = 1 \,.$$

In the interest of elegance and abbreviation, "a + (–a)" is usually written "a – a". This treats the sign " – " for the adjective "negative" as if it stood for the notion of subtracting: testimony, really, to Ohm's dynamic view that these are axioms for the operations as well as for what they operate on. It was this spirit that animated Newton, a century and a half before, not to ask futile questions about what gravity *is*, but to describe how it *acts* (its form, that is, rather than its substance). Masses and forces were on a par, as now were numbers and the forces on them.

The lawyers of mathematics tend to be satisfied with this list, but their clerks may insist on a prefatory pair to insure that the boiler-plate language is impenetrable:

> *The axiom of closure under addition*: If a and b are numbers, so is a + b.

> *The axiom of closure under multiplication*: If a and b are numbers, so is a · b.

Had you supposed that adding two numbers would produce a caterpillar, or multiplying them, a butterfly? Such axioms verge on mere definition—almost beneath the dignity of self-evidence.

Schumann wasn't alone in finding how shadowy objects become when their forms are separated from them. You might well ask at this point, "What are these the axioms *of*? Are these 'numbers' nouns or adjectives or verbs? Are they processes or the products which processes yield?"

One thing is certainly clear: not all of these laws have been brought from the inner sanctum of the natural numbers to the kingdom's extremes, as the impulse described on page 33 suggested. Consult again your New Pythagorean talisman on page 28: the axiom of additive inverses holds not in $\mathbb{N}$ but first in $\mathbb{Z}$; the axiom of multiplicative inverses not even in $\mathbb{Z}$ but first in $\mathbb{Q}$. It is as if the long revolution had moved the centers of power and interest out to the colonies, and the whole was now being ruled from them.

In truly formalist style this collection of axioms wasn't addressed to one kind of number or another but thought of by its first formulators as characterizing a self-standing whole: a *body* (*Körper*) as the German mathematician Richard Dedekind tellingly called it. Schumann might have brooded over this slight to any indwelling spirit, Pascal over the missing will.

What was a body in German became the even less suggestive "field" in English, the formalist point of view being that here was a list of laws and whatever obeyed them—rational, real, or complex numbers, or motions like the rotation of figures on a geometric plane, or chairs or beer mugs—was a field (the lure of abstraction may make mathematicians seem like a subspecies on the verge of evolving beyond lives steeped in the senses). Further relations could be deduced from those fundamental ones, and other relations could be shown *not* to hold among whatever obeyed them. It was the triumph of the container over the content: the slots stood to one another in specific ways; hence, so must anything slotted into them.

To take these laws in all at once, in a continuous sweep as Descartes would have us do, here are the unbroken Tablets of the Law, as delivered to us in 1893 by the equally abstract Heinrich Weber (a man about whom much is, but little more need be, known). They are expressed, only for convenience, in terms of numbers (a pure Formalist would have said: "If a, b, and c are elements of the field", and so on).

The Axioms for a Field
if a, b, and c are numbers, then

Under Addition		Under Multiplication
A0 a + b is a number	*Closure*	M0 a · b is a number
A1 a + (b + c) = (a + b) + c	*Associativity*	M1 a · (b · c) = (a · b) · c
A2 a + b = b + a	*Commutivity*	M2 a · b = b · a
A3 there is a number, 0, such that a + 0 = 0	*Identity*	M3 there is a number, 1, such that a · 1 = a; and $1 \neq 0$.
A4 for any number a there is a number, –a, such that a + (–a) = 0	*Inverse*	M4 for any number a, except 0, there is a number, $\frac{1}{a}$, such that $a \cdot \frac{1}{a} = 1$

D *Distributivity*: $a \cdot (b + c) = a \cdot b + a \cdot c$.

You may feel a need now for the axioms of subtraction and division—but see with what Spartan economy they have been included. Subtraction isn't a primary operation but is the inverse of addition; division, similarly, is just the inverse of multiplication. Their respective axioms let you balance the number line around 0 or 1.

We can also answer what seemed a merely rhetorical question in Chapter One. We asked on page 7: "multiplication is just sophisticated addition, isn't it?" The answer is: No. Certainly 3 · 4 means 4 added to itself 3 times, or 3 added to itself 4 times; but what does $3 \cdot \sqrt{2}$ mean? Three copies of $\sqrt{2}$ added together. The commutative law helps you make some sense of "$\sqrt{2}$ copies of 3 added together," but how could you explain at all in terms of addition what $\sqrt{7} \cdot \sqrt{2}$ means? Addition and multiplication are equally fundamental operations—Romulus and Remus (and commonly suckled by the Distributive Axiom)—but ultimately independent.

You might complain: "Where have you gotten this $\sqrt{2}$ and that $\sqrt{7}$ from? Since a field axiom gave us 1, another axiom produces 2, 3, and all the naturals, another their negatives, and a third the rationals—but nothing on the list accounts for the irrationals." And you are right so to complain: we need some way to assert their existence, and merely invoking different kinds of roots won't do, since as you saw on page 23 we can make irrationals in so many other ways.

Much energy and imagination, much argument and ink were spent on shaping something adequate and elegant enough to round out the table. In the background moved Schumann-like shadows that separated appearance ever further from being: for it is a trait of romantic enterprises that proxies beget proxies and what was stood in for turns out itself to have been a stand-in. So numbers gradually came to be thought of as secondary phenomena and *sets* emerged as fundamental. These, at last, had no antecedent and needed no definition. Unlike different sorts of numbers, we grasped sets at once and might call them "collections" or picture them as bags containing distinct objects, but this was mere paraphrase of what we knew without knowing (they weren't defined in terms of numbers or anything else, but now numbers could be defined in terms of them). Sets and their doings put bedrock under what had been shifting sands.

As early as 1835 the Irish mathematician William Rowan Hamilton—chaotic in life, discoverer of unguessed-at order in thought—came up with the idea that an irrational could be pinned down by dividing the rational numbers into distinct sets; and this idea Richard Dedekind brought to fulfillment later: he noted the date carefully in his diary (November 24, 1858), startled, perhaps, that a universal truth should enter human experience on a winter evening in Switzerland; or struck by his boldness at invoking completed infinities, since "set" snaps up, as a conceptual whole, an infinite array as swiftly as it does three buttons.

Richard Dedekind (1831–1916)

His idea was that if you break up the line of rationals into two distinct sets, with all those in the first to the left of each one in the second, the fracture between them is itself a number. Make the left-hand set all the rationals less than 2, for example, and the right-hand set all rationals greater than 2. Clearly 2 (which is a rational number: $\frac{2}{1}$) is the line of division. Instead fill the left-hand set now with all the negatives along with all the rationals whose *squares* are less than 2, and the right-hand set with all whose squares are greater than 2. Again there is a split between these two *infinite* sets: $\sqrt{2}$. Confer on this cut, says Dedekind, the status of number. Defining all the reals via these "Dedekind Cuts" breathes life into the irrationals among them, and the spirit of our laws is in that breath. For all that Dedekind's definition lacks the immediacy of our other terms, being as far above them in sophistication (using as it does infinite sets) as the closure axioms were below, it wonderfully catches the character of the irrationals, sifting like viruses through ever finer filters. As with all the best mathematical ideas, it catches too—and refines—a common way of thinking: "I don't know where the line is, but I know this was over the line," said a Boston detective in a case of child abuse.

The rationals, $\mathbb{Q}$, are a field, and enriched by Dedekind Cuts, so are the reals, $\mathbb{R}$. Add in $\sqrt{-1}$ and we find ourselves in the broadest field of the complex numbers, $\mathbb{C}$. We know the ancient name of the field this wall encloses: Eden.

$$\infty$$

Eden

For now we have our axioms, and logic enough to water them. Austere as these axioms seem, planted in a landscape as stark as biblical narrative, the vast and colorful garden of mathematics will grow from their seeds. You may wonder why we haven't added to our axioms the fact that $a \cdot 0 = 0$ for any a. The answer is: because we can now prove it, beginning to construct the tower of mathematics upward from the smallest possible base.

We start with $1 - 1$. The Additive Inverse Axiom, A4, assures us that this is 0. Multiply both sides of $0 = 1 - 1$ by any number a:

$$a \cdot 0 = a \cdot (1 - 1) \, .$$

Now apply the Distributive Axiom, D, to this equation's left-hand side, and you find that

$$a \cdot (1 - 1) = a - a \, .$$

But once again, by A4, $a - a = 0$. Shake these steps out in the right order and you get:

$$a \cdot 0 = a \cdot (1 - 1) = a - a = 0 \, ,$$

so that crossing the bridge of equalities from left to right gives us what we want:

$$a \cdot 0 = 0 \, .$$

This is an appetizer: a foretaste of the proving art. In it you won't catch a whiff of the kitchen from which it came: the cook's instinct for where to begin, the *mise en place* of the axioms and then their adroit use. There is a touch of the showman in mathematical presentations, where the deductions are made to look effortless.

Take, for example, a wild question whose answer nevertheless follows from these axioms. What are *all* the solutions to the equation

$$y^2 + y = x^3 - x \, ?$$

A sketch of the answer begins to materialize by giving specific values to x and then, using techniques derived from the axioms, finding what y's will produce these values. So for $x = 1$, y will have to be 0; and for $x = \sqrt[3]{5}$, we have $y^2 + y = 5 - \sqrt[3]{5}$ or $y^2 + y - 5 + \sqrt[3]{5} = 0$, and the y's turn out to be roughly 1.38 and −2.38. This sketch, filled in by yet further descendants of the axioms, shows that the infinite number of solutions lie on a "cubic curve", which when plotted on the coordinate plane has this curiously disjoint shape:

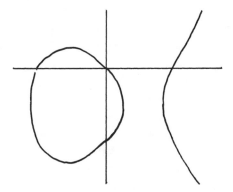

There you see all the real solutions to our equation.

But the Garden of Eden is famous for its snake, and the snake is the desire for more precise knowledge. Are there specific pairs of *integers*, those uniquely fundamental, ancient numbers, which satisfy this equation? Looked at again, you see that it can be rewritten as

$$y \cdot (y + 1) = (x - 1) \cdot x \cdot (x + 1) \, ,$$

so that with integers in mind we are asking: are there any numbers which can be expressed as a product of *two consecutive integers* (y and y + 1) and at the same time as a product of *three* consecutive integers (x − 1, x, and x + 1)? Not only has the question developed a profounder character when posed in terms of integers, but strangely enough, the axioms for the reals can't tell us enough to answer it! It isn't even clear, at first glance, how many integral solutions there are.

And yet, this isn't so strange: the axioms describe (or prescribe) the general life of the *reals*, not the specific mores of those living in $\mathbb{Z}$, close to the inner city of $\mathbb{N}$. Other means, attuned to these cloistered citizens, will have to be derived from their particular traits.*

Our understanding of the Pythagorean talisman has all at once to be turned inside out. You could no more expect to govern the founding city from the provinces than the Emperor Postumus could have hoped to govern Rome from Trier. The rationals, the reals, the complex numbers no longer appear as successive approximations to what ultimately is, but as ever more tenuous fictions flung in support around the central keep.

Perhaps this is what Kronecker meant when he said that God made the natural numbers and the rest was the work of man. You might even hear in this an echo of ancient Democritus: "By convention sweet," he said, "by convention bitter; by convention hot, by convention cold, by convention color: but in reality, atoms and void."

What are these other means for understanding the natural numbers? Rephrased in the Formalist style, what axioms describe them? The ingenious idea for proving properties peculiar to $\mathbb{N}$ is called *Induction*. Not children's induction: "Anything I say three times is true"; nor the peculiar Roman legal procedure called *ampliatio,* where an undecided jury could demand that all the evidence be repeated over (and over) again. Not even the sort of induction used in science, which concludes from a lot of test cases that a hypothesis *probably* holds. Certainty is the outcome here, and from many more than three or even a lot of instances: in fact, from *all* of them.

Inductive proofs work for statements about the natural numbers, by sweeping through their array as if they stood like dominoes, each less than a domino's length away from the next. Push over the first and all will tumble down. For the ingenious idea behind induction is this: prove that the statement in question is true for the first natural number it applies to—typically 1 (that corresponds to knocking over the first domino). Then show that *if the statement is true for any natural number, it must also be true for the next one* (that's the equivalent of checking that the dominoes are close enough together to communicate the initial impulse to all).

---

*The fact, for example, that there are exactly *ten* pairs of integers (x, y) that satisfy this equation is a hard-won twentieth-century surprise. You might have guessed that one pair is $x = 0$ and $y = 0$ ($0 \cdot (0 + 1) = (0 - 1) \cdot 0 \cdot (0 + 1)$) and another is $x = 1$ and $y = -1$: $((-1) \cdot (-1 + 1) = (1 - 1) \cdot (1) \cdot (1 + 1))$. But would you have come up with $x = 6$ and $y = 14$: that is, $5 \cdot 6 \cdot 7 = 14 \cdot 15$?

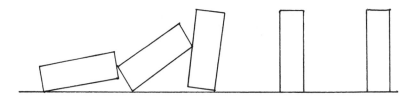

It is this second step that sets the mind's teeth on edge, since it looks as if we were assuming what we wanted to prove. Not so: we assume only that our statement is true for some number n, and then *using that assumption*, strive to show it must be true for the next, n + 1. If we succeed, then since the statement was proved to be true for 1, it must also be true for 2; but true for 2 must mean it is true for 3; and so 4; and therefore 5—*ad infinitum.* This is seeing the world in two grains of sand.

An example will help. If you like, you may then adopt medical school practice in mastering an operation: watch one, do one, teach one—a kind of human induction.

Here is the proof by induction of our already secure conviction that

$$1+2+3+ \ldots +n = \frac{n \cdot (n+1)}{2} .$$

First we establish that the claim is true for n = 1. Yes,

$$1 = \frac{1 \cdot 2}{2} .$$

Now assume it is true for any natural number—call it "a" for "any". We're assuming, that is, that

$$1+2+3+ \ldots +a = \frac{a \cdot (a+1)}{2} .$$

Using only logic, the bridges of equality, a few of our axioms, and this assumption, we now want to prove that the claim is true for the next natural number, which is a + 1—in other words, that

$$1+2+3+ \ldots +a+(a+1) = \frac{(a+1) \cdot (a+2)}{2}$$

(a + 2 is the successor of a + 1: (a + 1) + 1 = (a + 2)).

By our inductive assumption we can rewrite the left-hand side as

$$\frac{a(a+1)}{2} + (a+1) .$$

Using the Distributive Law, we can take the common factor $(a + 1)$ out of these two terms, giving us

$$(a+1)\cdot\left(\frac{a}{2}+1\right),$$

which is the same as

$$(a+1)\cdot\left[\frac{(a+2)}{2}\right].$$

That is,

$$(a+1)\cdot\frac{(a+2)}{2},$$

as desired. Magical, but watertight. And like a good piece of magic, the proof doesn't show us *why* the statement is true (as our two visual proofs did), only *that* it is so.

This is peculiar. Induction has confirmed the truth of many an important mathematical insight, but that insight had to have come from some other source. What induction does, in effect, is show that the insight spreads contagiously from a first number to the rest of the naturals, by making the insight clamp a number and its successor (n and $n + 1$) together: "the empty form", as it was called by a troubling figure of twentieth-century mathematics, the Dutchman L. E. J. Brouwer. For him, this is the form that remains when all the color is bleached from Before and After: the form of Induction that comes from recognizing that $1 + 1$ is a new whole (Brouwer calls it "two-ity").

Definitions like that about Dedekind Cuts might be hit on at a time and in a place, but we tend to think of methods (and certainly one this abstract) as timelessly there: part of our make-up. So in hefting a neolithic hand-ax and feeling it slip easily into your grasp, you think: "Of course—they made and used tools then as we do now." Patterns of use are immemorial. Yet induction too was invented by an embodied someone, not a figure as abstract as the empty form he dealt with. Francesco Maurolico was a Benedictine monk in sixteenth-century Sicily. Well, you think, the contemplative life would suit such abstract thoughts. Not a bit of it. He was head of the Mint; he was in charge of the fortifications at Messina; he devised various ways for measuring the circumference of the earth. He studied music, optics, magnetism, and the varieties of Sicilian fish; he successfully predicted for John of Austria what the weather would be like on the day of the Battle of Lepanto; he wrote a history of Sicily; he first observed the supernova that Tycho Brahe got credit for; in his spare time he translated Euclid, Apollonius,

Archimedes—and in the midst of this one thing after another of a life, he came up with Induction. When you look at the full moon you may see his memorial: the crater Maurolicus is named after him (dead scientists tend to become lunar and planetary features, dead mathematicians e-mail servers).

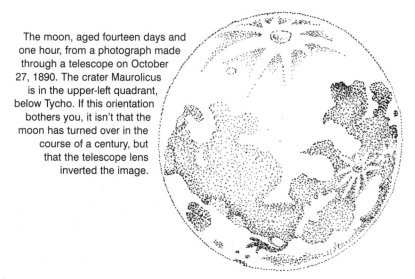

The moon, aged fourteen days and one hour, from a photograph made through a telescope on October 27, 1890. The crater Maurolicus is in the upper-left quadrant, below Tycho. If this orientation bothers you, it isn't that the moon has turned over in the course of a century, but that the telescope lens inverted the image.

In his book *Arithmetica*, Maurolico proves by induction that the sum of the first n odd numbers is n^2: the theorem whose truth we constructed on page 13. If you care to try an inductive proof yourself, remember the tripartite form:

1. Prove the statement true for n = 1;
2. Assume that it is true for n = k, and then
3. Prove that it is true for k's successor, k + 1.

You will probably also want to use the fact that the kth odd number has the form 2k − 1 and the next one 2k + 1 (you can check your proof against that in the Appendix).

It tells you how risky this new kind of proof must have seemed to its inventor that he actually checked the statement not only for 1 but for 3 and 5. It is as if we were witnessing scientific induction turning into mathematical induction.

History isn't inductive, since there never seems to have been a definitive first instance of any notion you can name. In the south of France, Levi ben Gerson, in 1321, used a process he called "rising step by step without end," which amounts to induction. Three centuries before him, Abu Bakr al-Karaji, in Baghdad, proved that cubing each of the first ten naturals and then taking the sum, was the same as the square of the sum of these naturals:

$$1^3 + 2^3 + \ldots + 10^3 = (1 + 2 + \ldots + 10)^2 .$$

His proof has an inductive air to it, working back, as he does, from the ten he wants to the truth that $1^3 = 1^2$ (the n = 1 case). He "must have" had induction in mind, just as Marie Antoinette "must have" thought longingly of Vienna just before the guillotine's blade fell. Arguments titillatingly close to induction appear around the same time in Ibn al-Haytham and al-Samaw'al (do they stop short with a sort of "et cetera", not seeing the pearl of price in their hands?)—and with hindsight we might even make out inductive reasoning descending the infinite staircase of past time, through and dimly beyond Euclid.

The clue that induction gave for axiomatizing the heart of the labyrinth, $\mathbb{N}$, was picked up by Dedekind and followed in 1889 by an Italian mathematician named Giuseppe Peano, who in the fervor of his purity sought to purge his language even of words and to use symbols instead: $\supset$ to mean "implies", $\in$ for "belongs to", and so on. His students rebelled. He tried to appease them by giving them all passing grades. It wouldn't do. He was forced to resign his professorship in Turin. The axioms, however, kept their chairs and are in them still. They are few in number, deceptively simple (may not all simplicity be deceptive?) and once again allowed "set" to loom as the undefined term behind the equally undefined "natural number", "successor", and "belong to". If God created the natural numbers, these were the words with which his prophet Peano, at least, began.

Peano's Axioms for the Natural Numbers

1. 1 is a natural number.
2. 1 is not the successor of any other natural number.
3. Each natural number n has a successor.
4. If the successor of n equals the successor of m, then n equals m.
5. [The key principle of induction] If a set S of natural numbers contains 1, and if n belonging to S implies that the successor of n belongs to S, then S contains all of the natural numbers.

You see, then, how "1" and "+", who walked hand in hand through the first chapter, now take their solitary way from Eden.

∞

The Mind and Truth

We spoke in Chapter One of the Alcibiades Humor in every mathematician's make-up: a brashness and daring that flouts convention and whose motto might well be "Why not?" Here you have seen a very different character developing: cautious, insistent on proof, neurotic with doubt. The wide gap between insight and demonstration might lead you to suspect that there were two sorts of mathematician, and you could tell from a glance at their monastic cells which were Jeromes with a lion, which Anthonys subduing demons.

It isn't like that. Riemann said "Just give me the insights. I can always come up with the proofs!" But his work is strung with diamonds of dazzling insights. The prolific eighteenth-century Swiss mathematician Leonhard Euler was renowned for his insights—but his proofs are gems. Both sorts live like two souls in the Faustian breast of each, not at war but in conversation with one another.

We could try distinguishing the halves of the mathematical brain along the lines of the character typologies you find in books with such charming titles as *On the Psychology of Military Incompetence* (impetuous, casual autocrats, riding to battle in old sweaters; obsessive, very clean authoritarians, sending their troops by timetable into the mud). What would we have?

Caricatures only, because of the immense variety of people who have prospered in mathematics—the reticent and the contentious, the companionable and the morose—and because this art is the birthright of us all, so deep in the structure of our thought that it is no respecter of origin or upbringing, of morality, age, even of sanity or madness. But since in mathematics we often make headway in a difficult problem by invoking extreme cases, let's vivify the opposite poles of mathematical activity—the drive toward insight, the urge to prove—by typifying each in men we have already met: the bitter rivals Brouwer and Hilbert. Each had profound insights, each produced stunning proofs—but Hilbert longed to establish mathematics on unshakable, impersonal foundations, Brouwer to free it from logic and even language, asserting the supremacy of private intuition.

For language, Brouwer wrote, "only touches the outside of an automaton." The soul, taken from the deepest home of its still center, deteriorates toward the external world through successive metamorphoses, in the last of which we merely enforce our wills on one another via language. Mathematicians generally believe that for all they invent their paths, the landscape through which those paths wander is *out there,* independent of them, its granite truths indifferent to their climbing. Brouwer turned away from such a view in disgust. These truths, he said, are "fascinating by their immovability but horrifying by their lifelessness, like stones from barren mountains of disconsolate infinity."

Instead, he saw mathematics as rooted in the mathematician's life and mind, which held the monopoly on certainty. Experience, axioms, logic itself had nothing to do with it, but intuition proceeded from primordial elements to a free and limitless unfolding. He dismissed any connection of intuition to the old Gnostic image of an inner light, or to any sort of Collective Mind: it was the individual mind that mattered—in fact, Brouwer's mind.

When he enrolled at the University of Amsterdam at sixteen, in 1897, he found himself surrounded by people who couldn't understand him and whom he couldn't stand. He made solitary pilgrimages to Italy, walking there and back in his large, dark cloak. He retreated again and again to his thatched hut in the forest, far from the world's motley plurality, to think mathematics directly, with eyes closed. For this was a game played in silence, as Brouwer's follower Hermann Weyl later put it. What the mind's eye saw were constructions fitting together. How reminiscent this is not only of the Stoic hand fitting the concept it grasped but also of the ancient Greek philosopher Xenophanes, who wrote that it wasn't fitting for God to move, but "without toil he shakes all things by the thought of his mind."

L. E. J. Brower (1881–1966). Like Descartes and Gauss, he had his best insights in bed.

By 1905 Brouwer realized that causal thinking fell into mere low cunning and was fundamentally immoral. Human nature was the real villain: mankind was like a bird arrogantly gulping up its own nest, as the Dutch had interfered with mother earth, gnawing and mutilating her with their dykes. Driven and aided by their characteristic human ability to reason, "some even start searching," he wrote, "for the foundations of their science . . . trouble increases and they go completely crazy. Some in the end quietly give up . . . they grow bald, short-sighted and fat, their stomachs stop working . . ."

He had a remarkable ability to attack positions he himself held, as though he had excluded any ground between contrary beliefs. A devotee of "air and mud baths", he wrote in 1905:

> . . . these inept people then take air baths, and when they discover the effect of sunrays they take to light baths and sun baths, and finally to dusk baths, night baths, moon baths, star baths, forest baths and meadow baths as soon as somebody proclaims them to be healthy.

Brouwer's troubles increased through his combative life as he failed to impose his will on his colleagues, lost his friends, and alienated all but his docile wife and a shadowy secretary. In the end, although he remained long-haired, lean, and fit, he quietly gave up promoting his revolutionary view of the way the mind builds up mathematics outside of language: a view each of us shares at the moment of insight. What Wordsworth wrote of Newton better fits Brouwer: his was

> . . . a mind for ever
> Voyaging through strange seas of Thought alone.

Meanwhile, what of Hilbert and his opposite desire to release mathematics from the vagaries of personality into universal acceptance? The field axioms made sense of what was built on them, but what made sense of the axioms? Doubt, as racking as a dry cough, shook the body of thought—for what did *self-evident* mean, after all? At best it implied some sort of circular reasoning; the alternative was a regress of justifications like the endless tower of turtles supposed to support the world—not an alluring infinity. And then, with the late nineteenth-century reduction of mathematics to set theory, cracks in the form of paradoxes began to open in the foundations themselves. The rot was everywhere. Doubts increased about Euclidean geometry's claim to the throne, with the pretenders from France, Russia, Hungary, and Germany brandishing their credentials. For each pictured space differently (Euclid's surfaces were flat, the others' differently curved), and nothing about them revealed which was the correct portrait of *actual* space. And there was the infinite. How could we, for example, accept Dedekind's claim that the irrationals really existed when, like anything in existence, they would have to take shape in finite time—yet lack of a pattern meant having to work out each of their infinitely many decimal places? Besides, his cuts required treating the sets that defined them as already completed infinities.

David Hilbert (1862–1943). From a set of postcards of the Mathematics Department sold by the University of Göttingen to tourists.

It would have been folly to beg these questions by shifting the burden onto some other subject, as if mathematics were descriptive of the world so that physics, say, or chemistry or the physiology of the brain would be ultimately responsible for its axioms—and in need therefore of axioms in its turn. Nor could doubting simply be dismissed: it was as necessary a component of the lust to know as aggression is of the sexual drive, though equally destructive in isolation.

But what about going back to the beginning of mathematics and revisiting and revising the Pythagorean vision? The harmony there could be rethought in terms of a harmony among the axioms. If they were *consistent* with one another, so that no paradox could follow from their workings, a rounded body of connections would grow musically from them. And if such axioms were also *complete*—sufficient, that is, for deciding the truth of every coherent statement these workings produced— then such a body would be a corpus of mathematics. Hilbert had already shown that the rival geometries stood or fell together: if one was consistent, they all were; if one harbored a contradiction, so did they all (see the appendix to this chapter for how he did this). He showed too that their consistency depended on that of the axioms for the natural numbers. The great project that he announced for the new century was therefore clear: logically to establish that *these* axioms were consistent and complete, so that arithmetic would stand in solitary splendor—and physics or astronomy could later sort out which of the equally plausible geometries happened to fit the observable facts.

How wonderful to collect together a harmonious set of foundations on which to build the edifice of mathematics. Nothing about this enterprise, however, guaranteed that the axioms would be an indubitable distillation of reality: that they were worthy of belief, as the word "axiom" meant to the Greeks. They would now be more like legal stipulations, or "postulates" in the Latin sense of agreements held merely for the sake of the argument. There would be no more hope of chasing them back to unquestionable truths than of following the etymology of "postulate" to anything less fanciful than petitioning kings or nominating bishops or wooing and praying.

To think of the axioms so is to recognize mathematics as a work of art, the free play of human thought; and the counters in this play, finite or infinite, would coexist in the harmony of consistency. But "play" is a word with a double lid, like a magician's hat: what goes in as the ace of spades comes out a rabbit. Being the playthings of the gods, said Plato, we should play the noblest games; yet games are what you are supposed to put away with adulthood. Didn't it trivialize our one contact with certainty to make its foundations just *coherent*—as whimsical as are the rules of Mah-Jongg? Wouldn't mathematics thus be reduced to no more than a meaningless system, the ultimate glass-bead game? Wasn't that where the Formalist impulse was driving us?

An equally frightening consequence lurked. If mathematics were shown to be a flawless machine that spun gears of definition past levers of logic, what need was there for any human to do more than set it in motion? The clockwork of 1900, the computer of 2000, could print out its theorems without our organic interference. Where then would be the glory of invention and the only claim our littleness has to the largeness of things?

The Intuitionists attacked. Without saying where the axioms came from, wrote the great French geometer, Poincaré, it would be as easy to postulate one as another, leaving our efforts arbitrary and incomplete. He himself suggested that our primary intuition was of mathematical induction. Brouwer put his finger on the key issue: existence. Just because a mathematical system was consistent, did that make it exist? "A false theory is false," said Brouwer, "even if not halted by a contradiction, just as a criminal act is criminal whether or not forbidden by a court." Only what thought can *construct* truly exists; and since what exists can't at the same time not exist, existence implies consistency. Hilbert's formal exercise might be ornate as a reliquary, but only the relic it housed— the shard of intuition—worked the miracles of mathematics.

Someone with a medieval turn of thought might have brought up what we'll call "The Great Converse" in Hilbert's defense. The medieval view was that creatures—the created—glorify God; so if there were more creatures, then the greater would be the glorification. Hence if something *could possibly* exist, it *would* exist. The world—as crowded with beings as the Unicorn Tapestry—would then more loudly sing God's praise. In Hilbert's terms this would translate to: since that which is consistent *can* exist, therefore it *must.* From this medieval standpoint, proving consistency would be enough to guarantee existence. Is it conceivable that Hilbert himself ever held this view? Could mathematical existence have meant this much—not this little—to him? He wrote to the profound logician Gottlob Frege on December 29, 1899:

> You write: "I call axioms propositions that are true but are not proved because our knowledge of them flows from a source very different from the logical source, a source which might be called spatial intuition. From the truth of the axioms it follows that they do not contradict each other." I found it very interesting to read this sentence in your letter, for as long as I have been thinking, writing, and lecturing on these things, I have been saying the exact opposite: If the arbitrarily given axioms do not contradict each other with all their consequences, then they are true and the things defined by the axioms exist. For me this is the criterion of truth and existence.

Of course you could see Hilbert as simply making sense of Peacock's Principle of Permanence. It didn't matter what the objects of mathematics were called or what the symbols stood for: the relations among them were the issue, and it was essential to guarantee that the axioms begetting those relations led to no contradictions. If the basement of your building is filling with water, you have to drain it before you can go back to furnishing the rooms. Once the foundations were secure, imagination and meaning, the play of thought and the freedom of mind, could return, and personality decorate the impersonal frame.

The arguments between Brouwer and Hilbert, at least, were soon drained of content and degenerated into mere form. Brouwer walked out of a dinner where Hilbert had been praised. Hilbert threw Brouwer off the board of the journal they both edited. They skirmished over who had published what first, and which mathematicians should go to a conference in Bologna, embittered as only angels who fall out can be.

∞

Brouwer's Silence and Hilbert's Laugh

Listening to them as amplifiers of the two voices in everyone who does mathematics, the message for us is important and complex. It has to do with the entangled roots of certainty and imagination. After a century such as the last you might cynically think that nothing indubitable was left, nor anything that couldn't be imagined. But imagination isn't fantasy: it means being able to focus ever more sharply on detail; and the wonder remains that the mathematics we conjure up turns out to describe the singular world we find ourselves in. We will therefore need to attend sympathetically to Hilbert and Brouwer, since it is from hardly explored reaches in ourselves that their voices speak.

Below their antipathy lay a surprising accord, and their lines of thought were more parallel than skew. As could happen in one of the new geometries (you will see why in Chapter Eight), these parallel paths coincided at their ends. Each began his career convinced that Kant was basically right: mathematics grew from intuitions shaping what our senses took in, even before we had any experiences—a shaping, in fact, that *allowed* us to experience. But the growing variety of geometries meant that Kant had to be wrong in some of the details. Gauss was the first to spot this: the variety must show, he said, that our knowledge of space turns out to depend on experience; but there are no competing systems of number because it alone comes from an intuition prior to experience.

Hilbert agreed. As part of his doctoral examination in 1885 he defended the *a priori* nature of *arithmetical* judgments (those, that is, about the natural numbers). Forty-five years later, in the farewell address he gave to his native city of Königsberg, he explained that after the dross had been removed from Kant's theory, "only that *a priori* will remain which also is the foundation of pure mathematical knowledge"—a foundation, he said, of intuitive insight.

William Rowan Hamilton, the Irish mathematician we met briefly before, wrote prophetically, as early as 1833, that the intuition of time was the sole source of number. With space now banished away by the purifiers, Kant's symmetrical foundations were broken, and the asymmetry took hold of the young Brouwer even more violently than it did Hilbert. He too defended the remaining bastion, but went beyond Kant, by turning intuition from passive stamp to active agent. He may never have read Hamilton, but like him declared that time was the primordial element from which mathematics came, and added that the Primordial Happening was becoming aware of one's existing in time. All of mathematics was made by detached and silent reflection on this Happening, during which the intuiting mind grasped something of time's reality. For this intuition yielded the two species of time: the "1" and "2" from "Then" and "Now" that gives us the natural numbers, and the continuous flow, from the ever in-between, that gives us the rationals and Brouwer's version of the reals, which for him exist only up through the decimal places we have finished constructing.

Despite what sounds like the ascetic calm of that detached reflection, Brouwer carried a world of capriccios darker than Goya's back to his hut. One had to be ever on guard against impurities that might creep into one's thought and nonintuitive germs that might be caught from others. And while he wrote publicly of "the fullest constructional beauty, the introspective beauty of mathematics," he jotted down in his notebook that "Mathematics and its application are sinful because of the intuition of time which is directly experienced as sinful."

He must, however, have been as blithe as Alcibiades in excluding contradictions from his moods, to plunge as he did with such unremitting vigor into the central task of his life. This was the need to prove his fundamental theorem. Since most of mathematics grows by using *functions* (think of them as rules or machines that effectively convert numerical inputs into numerical outputs), Brouwer had to show that functions worked accurately on numbers as he conceived them: built up in time. Natural numbers and rationals come well packaged, but irrationals, as you saw on pages 22 and 23, trail off fizzing and sputtering like the gauzy tails of comets. Brouwer's irrationals have two parts: the finite number of places we have made is like the comet's bright head; the rest is the tail we see fading away. His fundamental theorem would prove that if such a comet-like number entered into a function, a specific and similarly comet-like number (with an equally finite head) would come out. He hoped to prove that the output would depend only on the input's safely constructed head.

But his hopes lie scattered like Melancholia's tools around his abandoned building. What proofs he tried from 1923 onward all failed. The silence into which he retreated, until he died in 1966, was broken only now and again by fresh announcements of old programs, each ending on the threshold of the theorem he couldn't prove.

Hilbert meanwhile had all but finished his proof that the formal system of arithmetic—Peano's axioms—was consistent and complete, when a young Austrian logician named Kurt Gödel conclusively showed in 1931 that no proof of consistency and completeness could ever be made *within* the system to which it referred, as Hilbert's was meant to—for any system which was strong enough to deal with the mathematics of the natural numbers. While such a proof might exist in a larger system containing the first, *its* consistency would in turn have to be proven in a yet larger—

> And the great fleas themselves, in turn,
> Have greater fleas to go on;
> While these again have greater still,
> And greater still, and so on.

There was no hope, then, of securing the axioms by Hilbert's clever outflanking maneuver. He and his rival had moved through time, intent on its expression, to have their paths ultimately coincide in failure. Thousands of years before, someone had written an inscription on the base of the Egyptian statue of Neith: "I am all that was, that is, and that shall be, and no man has lifted my veil."

Was their failure inevitable because they had risen past the air that mind must breathe? Or had their counter-strivings managed to raise mathematics to a higher level? In his last publicly spoken words—a recording of the 1931 radio broadcast still exists—Hilbert said there could be no such thing in mathematics as an unsolvable problem. *"Wir müssen wissen. Wir werden wissen,"* he concluded: we must know. We shall know. And his biographer says that if you listen closely to the recording you will hear, through the crackle, a faint sound: Hilbert laughing.

The tension between Brouwer and Hilbert draws large the contrary poles in all who practice this art, showing mathematics's tensile strength—just as the weight of its compiled conclusions shows its compressive. Their struggle reached its climax between the wars, but we are always between two wars. The clamor over the foundations of mathematics only sleeps.* In this meantime no one would ask for justification by faith—

---

*Looking back over the rhythms in this history you might think that we shuttled between a few opposed positions. But the course of thought seems more of a spiral than a circle: new experiences and insights return us always above the positions we held and abandoned.

but might not justification by works serve? For when in the next chapters we look at the theorems built on those foundations, we will see radiant design and darting inspiration, the elegance of symmetry and asymmetrical surprises, preludial playfulness and fugal solemnity. Will this not be convincing evidence that the tower of mathematics is firmly founded?

Designs on a Locked Chest

Languages grew confused as the tower of Babel rose—perhaps because its foundation in all the variety of a common speech was too broad. The tower of mathematics is inverted, widening up and outward from its few axioms. These unify a greater and greater diversity.

Having walked one turn up the spiral, we are now where the immediate consequences of the axioms live. Here the first questions that the axioms give rise to can be answered. Is 0 the only additive identity and 1 the unique identity for multiplication? A proof as firm as a handshake shows this is so, along with another, that each number has only one inverse (the proof isn't in the eating but the Appendix). It is on this turn that the proof you have already seen, that a · 0 = 0, is at home; and it is here that the row about dividing by 0 is settled.

"I can more easily imagine cutting something up zero times than I can a million times!"

"Only God can divide by zero."

"You can multiply by zero, and division is paired with multiplication, so you can divide by zero."

The keeper of the axioms comes out of his shop holding his favorite instrument, the proof by contradiction.

"Assume," he says, "that you could indeed divide by 0. That means 0 has a multiplicative inverse, just like any other number. My axiom M4, about multiplicative inverses, guarantees that

$$\frac{n \cdot 1}{n} = 1 \, ,$$

so if n = 0,

$$0 \cdot \frac{1}{0} = 1 \, .$$

"Now choose a number, and I'll choose a different one; I'll call yours a and mine b. So a ≠ b. If you feel that a card was forced on you, remember my third axiom of multiplication. It said that there were indeed at least two different numbers: 1 ≠ 0; or if you'd rather align yourself with Brouwer, 1 is different from 2.

"And now we will reach our goal together, in the way of mathematics, by building on previous work. We know that a · 0 = 0 and b · 0 = 0, so

$$a \cdot 0 = 0 = b \cdot 0 \,,$$

hence

$$a \cdot 0 = b \cdot 0 \,.$$

"Multiply each side of this equation by the supposed inverse of 0, $\frac{1}{0}$:

$$(a \cdot 0) \cdot \frac{1}{0} = (b \cdot 0) \cdot \frac{1}{0} \,.$$

Now use the Associative Axiom of Multiplication to shift the parentheses:

$$a \cdot \left(0 \cdot \frac{1}{0} \right) = b \cdot \left(0 \cdot \frac{1}{0} \right) \,.$$

But we saw that $0 \cdot \frac{1}{0}$ would have to be 1, so what we have is

$$a \cdot 1 = b \cdot 1 \,,$$

in other words,

$$a = b \,;$$

and this contradicts a being different from b. Since the logic of each step was sound, the only thing that could have gone wrong was assuming that $\frac{1}{0}$ exists—that is, that you could divide by 0."

This sort of proof, like close-up magic, is over so quickly that it takes walking slowly around it to convince yourself of its legitimacy and importance. The arguments over the centuries about whether $\frac{0}{0}$ was 0 or 1, indefinite or infinite, have been settled in a moment: none of the above.

Farther along this first tier lies the answer to a puzzle that has put too many people off math forever, convinced that its dicta were arbitrary or spiteful: for why should the product of two negative numbers be positive? We won't be helped by remarking that two wrongs don't make a right, and we can't get no satisfaction from double negations being assertions. J. B. Brown, writing in a 1936 *Punch*, speaks for us all:

Long ago, when a small scrubby schoolboy
A mixture of Etons and ink
(*Eheu fugaces*! How time simply races!
Said somebody, Horace, I think),
Whatever the lesson I read it and said it
Without the least trouble or fuss—
But I never could see how on earth it could be
That – *plus* – made –
But – *times* – made +

So, just as the ancient Achilles
Had his heel for opponents to pink,
The armour, it's plain, of my versatile brain
Has its single assailable chink.
Must I always in ignorance wander? I ponder—
For ever be limited thus?
Or will it be clear ere I vanish from here
Why – *plus* – is –
But – *times* – is +?

Mathematics is synthetic, so again let's build on our recent gains. Instead of a proof by contradiction, we will spin out from the axioms a web fit to catch this mystery. If a and b are positive numbers, we want to know whether (–a) · (–b) is negative, positive, or what.

We certainly sense that a *positive* times a negative is negative: owing 4 people $3 puts you $12 in debt, and the Commutative Law assures us that 4 · (–3) is the same as (–3) · 4: a negative times a positive is negative. You'll find a more rigorous proof in the Appendix.

We are now ready to learn what (–a) · (–b) is. By the Additive Inverse Axiom, b – b = 0. Multiply both sides of this equation by –a:

$$(-a) \cdot (b - b) = (-a) \cdot 0 \,.$$

Since anything times zero is zero, this simplifies to

$$(-a) \cdot (b - b) = 0 \,.$$

Ignore the fact that (b – b) is 0: it won't get us anywhere. Instead, look at the equation from the different angle provided by the Distributive Axiom; then it turns into

$$(-a) \cdot b + (-a) \cdot (-b) = 0 \,.$$

From our previous discussion, we know that $(-a) \cdot b$ is $-(a \cdot b)$, so our latest equation can be rewritten as

$$-(a \cdot b) + (-a) \cdot (-b) = 0 \,.$$

Pause for a moment to consider this. $-(a \cdot b)$ plus something—our mystery guest—is zero. But we already know what you have to add to $-(a \cdot b)$ in order to get zero: its additive inverse, the positive number $a \cdot b$! So

$(-a) \cdot (-b)$ is $a \cdot b$ in disguise:

$$(-a) \cdot (-b) = a \cdot b \,.$$

The product of our two negatives is positive, no matter how happy and undeserving a may appear, or how wretched and meritorious b.

We can only hope that this became clear to J. B. Brown ere he vanished from here, and are grateful that the axioms have let us prove the truth of what confounded our intuition. Have negative numbers definitively moved mathematical thought into abstraction, where the dance of symbols becomes vivid instead of figures? Or do you find the visual proof in the appendix to this chapter not only convincing but illuminating? Notice that in our dances the same steps—axioms of additive and multiplicative inverses, and distributivity—occur again and again. This is because, like squaredances in the confines of a barn, little room to maneuver leads to intricate patterns. The more elaborate these become, each linking onto the last, the more such patterns will all seem to lodge in a sense at once more ancestral and more abstract than sight. It is as if the predominance in our brains of the visual cortex masked a different, deeper apprehension—of time, then, or something akin to music: structure itself.

∞

The Primal Secret

Whatever we find as we spiral up the tower appears to be made by addition or multiplication. Addition you know like the back of your hand. Its axioms wholly describe it and its basic building block is the number 1. Since the axioms for multiplication are all but the same as those for addition, shouldn't you know it as well as your palm? The two operations have different identity elements, 0 and 1; the roles they play in the distributive axiom differ—and that's all. But what are the building blocks of multiplication? The *composite* number 24 breaks down into a sum of 1s, but a product of what—2 and 12, or 3 and 8, or 4 and 6? If you look for the atomic factors of 24 you end up with

$$2 \cdot 2 \cdot 2 \cdot 3,$$

or $2^3 \cdot 3^1$. These factors are the atoms of multiplication because no further factor (save the anonymous 1) divides them.

They are called the *prime numbers,* and simple as they are to describe, nothing in all of mathematics has turned out to resist more stubbornly our efforts at understanding. Here we are, wholly within the circle of the natural numbers, where it is equally natural to ask: what pattern is there to the primes, and how do they behave? Insights into these questions are so few and far between that each is celebrated as a major victory, and people otherwise lost to time are remembered for one telling conjecture about them.

Even the most elementary question brings us up short: how many primes are there? The first few are

$$2, 3, 5, 7, 11, 13, 17, 19, 23, 29, 31 \dots *$$

Looking at that line-up and consulting the inner oracle of your intuition, you might be tempted to think that the primes went on forever, since the natural numbers do, and these are among them. Or you might reason that the farther out you go, the fewer there will be, since the larger the number, the more factors pile up behind it—so after some point the primes might simply give out. This argument is supported by a preliminary survey: there are 25 primes between 1 and 100, 6 between 1,000,000 and 1,000,100, but only 2 between 10,000,000 and 10,000,100. We are wise enough in the vagaries of numbers by now, however, to sample only as a geologist studies a pebble: not to deduce the mountain from it, but to build up the evidence in his trays.

The very finite proof that in fact the number of primes is infinite stands framed in Euclid—so it is at least 2,300 years old. It needs only the briefest introduction. When we say "divides" in this chapter on Number Theory we will always mean "without a remainder"—so 3 divides 9 but not 10. Now we are prepared to follow, as if they were our own, the thoughts of a mind so very far from ours in time and context.

---

*We omit 1 because although it fits the definition of "prime"—a natural number divisible just by itself and 1—it would only clutter up the uniqueness of prime factorization: 24 could be written as $1 \cdot 2^3 \cdot 3^1$ or $1 \cdot 1 \cdot 1 \cdot 1 \cdot 2^3 \cdot 3^1$ or any number of useless 1s scattered among the substantial factors of 2 and 3.

Euclid wants to prove that there is no last prime. He does this by show-ing that no matter how many primes you have, you are forced to pro-duce another. For multiply together all the primes you can think of in order, stopping at some prime p, and call the product of them all n:

$$2 \cdot 3 \cdot 5 \cdot 7 \cdot 11 \cdot 13 \cdot \ldots \cdot p = n \, .$$

Clearly, every single one of these primes divides n.

But if we have n, we must also have n + 1. This seriously large number is very much greater than any prime in the collection, but it can't be divided by any of the primes from 2 to p, since division by any of them would leave a remainder of 1.

You would like to conclude now, with glee, that n + 1 must therefore be prime—but we have to attend first to a small point of order. n + 1 might not have any of the primes from 2 to p as a factor, yet it might still be composite if it had a prime factor q somewhere in the great gulf be-tween p and n + 1. Well and good: then q would be the new prime.

This flash of a proof lights up the infinite vista of natural numbers enough for us to see that the primes in their niches are stationed end-lessly there. Its beauty lies not only in the beam's pure light, but in achiev-ing so much with so little.

The twentieth-century mathematician Paul Erdos often spoke of "The Book": the book, he meant, in which God keeps all the most beautiful proofs. "You don't have to believe in God," said Erdos, "but you do have to believe in The Book." Everyone has his own edition of this book, but Euclid's proof of the infinitude of primes is likely to be in them all.

After such a breakthrough you would ex-pect hordes of results about the primes to start pouring in. We know that there are infinitely many multiples of 3, and if asked for a for-mula which would give us any particular one, such as the eighteenth, could do so with ease (54), by way of the expression 3n. Yet even this we still can't do for the primes. A clever Greek named Erastosthenes, in the third century B.C., did make use of such patterns to sieve out the primes in a purely mechanical way—not by a formula but from what all formulas like 3n left behind. This is the way it worked.

Write out the natural numbers from 2 on for as long as you like, then cross out the mul-tiples of 2, then of 3, then of 5—but leave 2, 3,

Erdos at eight. The book in his hand is most likely not yet The Book.

and 5 themselves standing. The next number you come to has to be a prime, so leave it in place but cross out all its multiples, and repeat the process. What you are left with will be just the primes scattered through your original table.

Before

2 3 4 5 6 7 8 9 10 11 12 13 14 15 16 17 18 19 20 21 22 23 24 25 26 27 28 29 30 31 32 33 34 35 36 37 38 39 40 41 42 43 44 45 46 47 48 49 50 51 52 53 54 55 56 57 58 59 60 61 62 63 64 65 66 67 68 69 70 71 72 73 74 75 76 77 78 79 80 81 82 83 84 85 86 87 88 89 90 91 92 93 94 95 96 97 98 99 100 101 102 103 104 105 106

During

2 3 4̸ 5 6̸ 7 8̸ 9̸ 1̸0̸ 11 1̸2̸ 13 1̸4̸ 1̸5̸ 1̸6̸ 17 1̸8̸ 19 2̸0̸ 2̸1̸ 2̸2̸ 23 24 2̸5̸ 2̸6̸ 2̸7̸ 2̸8̸ 29 3̸0̸ 31 3̸2̸ 3̸3̸ 34 3̸5̸ 3̸6̸ 37 3̸8̸ 3̸9̸ 4̸0̸ 41 4̸2̸ 43 4̸4̸ 4̸5̸ 4̸6̸ 47 4̸8̸ 4̸9̸ 5̸0̸ 5̸1̸ 5̸2̸ 53 5̸4̸ 5̸5̸ 5̸6̸ 5̸7̸ 5̸8̸ 59 6̸0̸ 61 6̸2̸ 6̸3̸ 64 6̸5̸ 6̸6̸ 67 6̸8̸ 6̸9̸ 7̸0̸ 71 7̸2̸ 73 7̸4̸ 7̸5̸ 7̸6̸ 7̸7̸ 7̸8̸ 79 8̸0̸ 8̸1̸ 8̸2̸ 83 8̸4̸ 8̸5̸ 8̸6̸ 8̸7̸ 8̸8̸ 89 9̸0̸ 9̸1̸ 9̸2̸ 9̸3̸ 94 9̸5̸ 9̸6̸ 97 9̸8̸ 9̸9̸ 1̸0̸0̸ 101 1̸0̸2̸ 103 1̸0̸4̸ 1̸0̸5̸ 1̸0̸6̸

After

2	3		5	7		11	13			17	19			23	
			29	31					37			41	43		
47					53				59	61					67
			71	73				79			83				
89							97			101		103			

Ingenious? Yes. A work of genius? No. Eratosthenes seems to have been the first person for whom that English put-down was used, "a Beta mind". His device would allow slaves then and computers now to spell out the primes, but without any insight into their structure—without even any need to know multiplication tables. Repeatedly counting up to three, up to five, up to seven, and so on suffices, and isn't to be despised: you will see in Chapter Nine how counting alone will open windows on a landscape more dramatic than any in fantasy fiction.

An enormous number of primes has been amassed since Eratosthenes's day, and our casual statistics on page 60 seem to show them dwindling away the farther along we go—yet now we can add: without ever disappearing. Perhaps if you laid out regular intervals you would find steadily fewer in each, like settlers in the first westward scatter past the Appalachians. To test this, let's look by hundreds at the stretch from 1 to 1000:

Between	The Number of Primes Is
1 and 100	25
100 and 200	21
200 and 300	16
300 and 400	16
400 and 500	17
500 and 600	14
600 and 700	16
700 and 800	14
800 and 900	15
900 and 1000	14

This is faintly disquieting: the number of primes bumps down and up as it declines. Perhaps it will even out as we move much farther along:

Between	The Number of Primes Is
1,000,000 and 1,000,100	6
1,000,100 and 1,000,200	10
1,000,200 and 1,000,300	8
1,000,300 and 1,000,400	8
1,000,400 and 1,000,500	7
1,000,500 and 1,000,600	7
1,000,600 and 1,000,700	10
1,000,700 and 1,000,800	5
1,000,800 and 1,000,900	6
1,000,900 and 1,001,000	8

Curiouser and curiouser. As bumpy as before, but at least the numbers are consistently lower. If we move up to the thousand-long stretch from 10^7, that is, 10,000,000, on, we might expect minor perturbations, but at least we won't see any interval with 10 primes in it again—or will we?

Between	The Number of Primes Is
10,000,000 and 10,000,100	2
10,000,100 and 10,000,200	6
10,000,200 and 10,000,300	6
10,000,300 and 10,000,400	6
10,000,400 and 10,000,500	5
10,000,500 and 10,000,600	4
10,000,600 and 10,000,700	7
10,000,700 and 10,000,800	10
10,000,800 and 10,000,900	9
10,000,900 and 10,001,000	6

The law governing the distribution of primes must be quite subtle—for surely there is some law. At any rate, we haven't found 25 primes in a span of 100 numbers this far out, or any of those concentrations we saw between 1 and 1000. Perhaps then the distribution of primes is settling down toward some constant number in any hundred-unit run—two, say, or three. We last looked at 10^7. By the time we accelerate away to the trillions, for example, we might expect fewer than 6 in any patch of 100. Disappointment again:

Between	The Number of Primes Is
10^{12} and $10^{12} + 100$	4
$10^{12} + 100$ and $10^{12} + 200$	6
$10^{12} + 200$ and $10^{12} + 300$	2
$10^{12} + 300$ and $10^{12} + 400$	4
$10^{12} + 400$ and $10^{12} + 500$	2
$10^{12} + 500$ and $10^{12} + 600$	4
$10^{12} + 600$ and $10^{12} + 700$	3
$10^{12} + 700$ and $10^{12} + 800$	5
$10^{12} + 800$ and $10^{12} + 900$	1
$10^{12} + 900$ and $10^{12} + 1000$	6

After having so triumphantly proved so long ago that there are infinitely many primes, why are we having such trouble in the twenty-first century answering this simple question about their distribution? Perhaps our approach has been wrong. Let's ask instead if we will ever find a gap larger than 6 between consecutive primes (6 was the gap between 23 and 29). There must be, since there is, for example, only one prime between $10^{12} + 800$ and $10^{12} + 900$, hence a gap of at least 50.

The startling news is that there are stretches of numbers a thousand long with not a single prime among them. More: there are primeless stretches a million long! Since the primes never end, you *will* come on one eventually after such a span—which begins to give a horrifying sense of how big very big numbers are, and how immeasurably bigger than big the infinity of the natural numbers is.

Yet we have hardly begun. There is at least one run of natural numbers a trillion long where there are no primes whatsoever; and another ten trillion long; and another—but you probably think that no human could possibly know this for sure, or that it would take an equally gigantic mind to understand it. In fact, the proof is at your fingertips, so perversely beautiful and innocently powerful is mathematics. This proof is a variation played on Euclid's theme.

We need a new symbol to help us, and the one commonly used is appropriate to the astonishment of our theorem that *there are strings of numbers as long as you like that haven't a single prime in them*. This symbol is !, called *factorial*. Written after a natural number it means: take the product of all the natural numbers up to this one. So $3! = 1 \cdot 2 \cdot 3 = 6$, $5! = 1 \cdot 2 \cdot 3 \cdot 4 \cdot 5 = 120$, and $n! = 1 \cdot 2 \cdot 3 \cdot 4 \cdot \ldots \cdot n$.

Now choose any number n you like—n could be 7 or 93 or 65,537—and make the much larger number n!. Consider the following string of consecutive numbers:

$$n! + 2, n! + 3, n! + 4, n! + 5, \ldots, n! + n .$$

We don't start with $n! + 1$, since it might be prime; but $n! + 2$ can't be, since 2 divides each part and hence is a factor of the whole. In the same way, 3 is a factor of the second number, 4 of the third—and so on, up to the last, of which n is a factor. So none of these numbers—there are $n - 1$ of them—can be prime! If you want, then, to make a string of numbers 78 octillion long which is guaranteed to be without a prime, let n be 78 octillion plus one ($78 \cdot 10^{27} + 1$ or 78,000,000,000,000,000,000,000,000,001), and the required numbers will run from (78 octillion + 1)! + 2 to (78 octillion + 1)! + (78 octillion + 1). After that prime-free run, somewhere, there will be another prime. . .

A proof like this makes you as giddy as does looking too long at the night sky, where darkness and stars unequally recede:

> This lonely hill was always dear to me,
> And this hedgerow, that hides so large a part
> Of the far sky-line from my view. Sitting and gazing,
> I fashion in my mind what lie beyond—
> Unearthly silences, and endless space,
> And very deepest quiet; then for a while
> The heart is not afraid. And when I hear
> The wind come blustering among the trees

I set that voice against this infinite silence:
And then I call to mind Eternity,
The ages that are dead, and the living present
And all the noise of it. And thus it is
In that immensity my thought is drowned:
And sweet to me the foundering in that sea.*

Just when you think you have come to terms with the inhuman dimensions of the primes, they deliver a shock with a shiver of intimacy in it. For if you go back to the list on page 60 of the first few primes, you'll notice that there are some which are only two apart: 3 and 5, 5 and 7, 11 and 13, 17 and 19. By now you would expect such pairs—"twin primes" as they're called—to thin out and disappear in the outer reaches. Yet 101 and 103 are both prime—and so are 809 and 811, and 3,119 and 3,121, 10,005,427 and 10,005,429. It has long been suspected that although they recede from one another like red-shifted stars, there is an infinite number of these twin primes. Like most of the questions in mathematics, this one has yet to be answered, and all we can do thus far is come up with new champions. The largest twin primes on record, to date, are $1,807,318,575 \cdot 2^{98,305} - 1$ and $1,807,318,575 \cdot 2^{98,305} + 1$. Each would take about thirteen of these pages to write out in full, and would likely make tedious reading. The excitement lies in the acrobatics needed to find them (a discovery made just as we were writing these pages. For the actual announcement, see the on-line Annex).

Where does that leave the primes? Irregular surfacings, gigantic gaps, occasional twins—we are so used to pattern coalescing at last from chaos (reading the geological record from strewn fossils) that we proudly think: where pattern is, our minds will find it. What if after all, then, there *is* no pattern, and multiplication, unlike addition, is built—built by us!—on utterly chaotic foundations? Could such slender differences in the axioms of addition and multiplication lead to divergences this profound?

Since there seems to be no rhythm to the primes, let's invert our way of looking and ask if different rhythms carry different quantities of them. A 3-rhythm starting with 3—

$$3, 6, 9, 12, \ldots$$

has only one prime: that initial 3. What about a 3-rhythm starting with 2?

$$2, 5, 8, 11, 14, 17, \ldots$$

where each number is one less than a member of the 3-times table—so the rule is that the nth term of the sequence is 3n − 1.

---

*Leopardi, "The Infinite", in John Heath-Stubbs's translation.

A proof, very like Euclid's for the infinite number of primes, will show that there is an infinity of primes even in this sequence. The proof slips on like a glove. Assume there are only finitely many primes in the sequence 2, 5, 8, 11, 14, 17, ... and so on, to a last one, p. Now multiply all these numbers together, multiply that product by 3, and then subtract one. Call this result m:

$$m = 3 \cdot (2 \cdot 5 \cdot 8 \cdot 11 \cdot 14 \cdot 17 \cdot \ldots \cdot p) - 1 .$$

m is clearly much larger than p, the supposedly last prime in the sequence, and m is *also* in our sequence, because it is of the form 3n − 1. Amazingly enough, any number of this form has at least one prime factor of the form 3n − 1 (consult the Appendix for a proof). Yet we know from our little discussion, back in the innocent days of page 61, that none of the primes up to p could be that factor, since each is already a factor of $3 \cdot (2 \cdot 5 \cdot 8 \cdot 11 \cdot 14 \cdot 17 \cdot \ldots \cdot p)$ and hence would leave a remainder of 2. So m is either prime and destroys p's haughty claim to be the largest prime of this form, or it has a prime factor somewhere in the gap between p and m, and this prime factor would also have the required form and be larger than p.

The story of how many primes there are in such sequences was finally told in 1837 by a remarkable man named Johann Peter Gustav Lejeune-Dirichlet (an ancestor, a young man from Richelet—"le jeune de Richelet"—moved from Belgium to the Rhineland. Naturalization changed the spelling over time but some ancient memory saw to it that, though the "ch" became hard, the "let" is still pronounced as it was in French). Johann married Mendelssohn's sister Rebecca, and since everything is connected to everything else, it should come as no surprise that Johann was taught by Georg Ohm, whom we met in the second chapter, and in turn taught Kronecker, whom we met in the first. What Dirichlet proved—one of the landmarks in number theory—was that *any* sequence of the form an + b will have infinitely many primes in it, as n goes from 1 through the natural numbers; all that is required of a and b is that they have no common factor.

Once again we are at a loss in trying to see the structure of the primes: no particular rhythm carries more of them than another. Yet if we assume chaos we cannot but deduce despair. Since intuition and common sense have left us stranded, we need an insight—and then a proof for it to nestle comfortably into. Gauss—whom we saw as a schoolboy triumphantly writing on his slate—used to contemplate tables of primes for sheer amusement, the way Russians always and the English on country house weekends love browsing through railway timetables.

He would while away spare hours calculating in his head which numbers were prime in runs a thousand long. This sort of rambling among the naturals, like a lepidopterist out with his net, was to gain him not only a collection of iridescent creatures but give him the basis, at last, for something approaching their taxonomy.

His intimacy with the raw data led him to mull over a question with a statistical flavor: ignoring the stuttering way they pop up, might there yet be some regularity in how the sheer number of primes increases? Let's graph how many primes there

Carl Friedrich Gauss (1777–1855), a mason's son and the master builder of mathematics.

are up to the number x (in our diagrams the horizontal axis will be the inputs: values for x; and the vertical axis the outputs: number of primes less than or equal to x). This function is commonly called π(x), meaning the number of primes less than or equal to x (that "π" has nothing to do with the π from geometry, but was chosen so that the Greek p would remind us of "prime"). So π(3) = 2, since there are two primes (2 and 3) less than or equal to 3, π(4) is also 2, and π(8) = 4 (2, 3, 5, 7 are the primes less than or equal to 8).

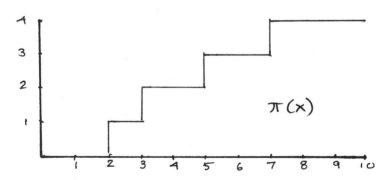

Only the horizontal "treads" matter in this step-graph. The vertical "risers" are conventionally put in just to give it a coherent shape.

Here is the graph of π(x) for x up to 100 (in order to accommodate the slow growth of the primes, we have shrunk the units on the vertical axis until those on the horizontal axis are about seven times their size, so that the graph looks much steeper than it should):

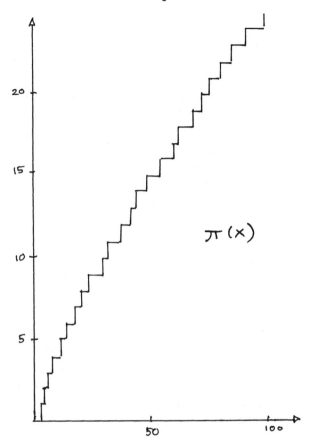

This rises by uneven steps, but on a staircase with at least a steady camber to it, so that a brush stroke would make the rough places smooth—and a smooth curve stands a good chance of representing a congenial function.

Now when you look at $\pi(x)$ for x up to 50,000, you see such a curve: the irregularities have startlingly been spirited away, as the remote full moon makes a perfect circle in the sky.

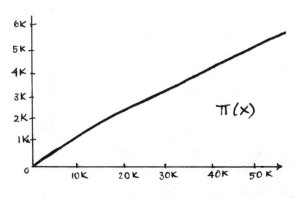

Yet smooth as the curve may be, we can't predict how it will continue, or understand what gives it the shape it has, unless we discover a function which accounts for it: a function whose graph it is.

The amazing fifteen-year-old Gauss came up with such a function. He looked at the data we saw on page 63 and realized that the ratio of x to $\pi(x)$ increased by roughly 2.3 from one power of 10 to the next. 2.3? To someone utterly engrossed in mathematics this number will ring with the familiarity that "To be or . . ." has for a reader of English: it is the beginning of a famous exponent. There is a number e—an irrational close to 2.7—which lies at the heart of biology and economics, because it expresses organic growth. When e is raised to about 2.3 you get 10. The eccentric Scottish mathematician John Napier cobbled together two Greek words, *logos* and *arithmos,* to make "logarithm", for talking about what exponent is needed to raise a chosen number (the base) to reach the number you want. Since "$2^3 = 8$" says you must raise the base 2 to the power 3 in order to get 8, Napier wrote: the logarithm with base 2 of 8 is 3 (abbreviated $\ln_2 8 = 3$). The number you need to raise e to, in order to get 10, is about 2.3 ($\ln_e 10 \approx 2.3$; most people simply write $\ln 10 \approx 2.3$, where "ln" by itself means with base e). A brief note in the Appendix explains e and its logarithm.

Gauss therefore leapt to the conjecture—on the basis of how the primes were distributed among the first 3,000,000 integers!—that $\pi(x)$ was closely followed by $\frac{x}{\ln x}$. You see here how well the two curves match:

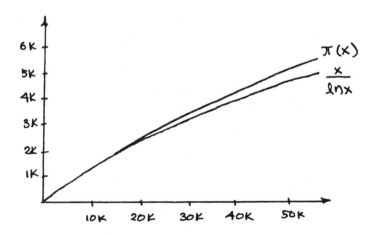

Logarithms and irrational numbers? How could these creatures of realms so remote from the naturals have any bearing on the primes? Perhaps because our looking is statistical; or because no sort of number is an island but each is a part of the main, and the sea of functions implicates each in all.*

Gauss was unable to prove his conjecture, which did not become a theorem until 1896, when it was proved independently in two very different ways by two very different men, who were born a year apart and died a year apart, almost a century later; their lives were to diverge radically from their intersection at the proof of this conjecture. One, who gloried in the name Charles Jean Gustave Nicolas de la Vallée Poussin, was born, lived, and died in Louvain, in Belgium: a professor, like his father before him, at the university there; survivor of two world wars and fifty years of teaching. The other, Jacques Hadamard, was good in all subjects but math when at school ("In arithmetic until the seventh grade I was last or nearly last"); worked vigorously to clear his relative Dreyfus; had two sons killed in the First World War, and fled from France to America during the second. Each sought a solution to the clarion call of this problem in the texture of ideas and techniques thickening around it, and followed his separate clue out of the labyrinth.

You will notice that the graph of $\frac{x}{\ln x}$ stays below $\pi(x)$ up to x = 50,000. Gauss, endlessly fecund, came up with an even better approximation to $\pi(x)$, this one narrowly overestimating it, up to at least x = 1,000,000,000. His new approximation, called Li(x), involved a notion at the heart of calculus, the integral:

$$\text{Li}(x) = \int_2^x \frac{1}{\ln t} dt \ .$$

Here the eighteenth-century elongated S denotes the *area* between the x-axis and the curve traced by the function (in this case that function is the reciprocal of the logarithm function) between two vertical lines (here set up at 2 and, to its right, at x).

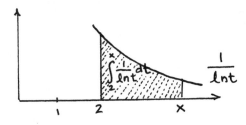

*e is by no means the only irrational that lives with the primes. The π of geometry, which is an irrational beginning 3.14159..., has pitched its tent in their midst. For if p_1, p_2, p_3 and so on are the primes in order, then the infinite *product*,

$$\frac{1}{1-\frac{1}{2^2}} \cdot \frac{1}{1-\frac{1}{3^2}} \cdot \frac{1}{1-\frac{1}{5^2}} \cdot \ldots = \frac{1}{1-\frac{1}{p_1^2}} \cdot \frac{1}{1-\frac{1}{p_2^2}} \cdot \frac{1}{1-\frac{1}{p_3^2}} \cdot \ldots = \frac{\pi^2}{6}$$

and π^2 is irrational too. Euler first miracled this up. We still hardly understand all it implies.

You see the remarkable accord between π(x) and Li(x) up to x = 50,000 (the area Li(x) measures, after all, grows as x grows, and like π(x), ever more slowly):

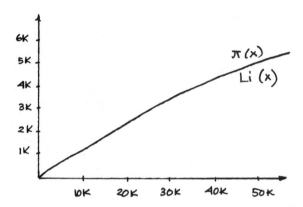

On this scale you can't even see the difference between them, yet up to very large x, Li(x) always overestimates π(x). Gauss remarked that for x = 400,000, π(x) = 33,859 and Li(x) ≈ 33,922.621995—a difference of less than .2%. Does it always overestimate? Strangely enough, no. Somewhere very far out, π(x) becomes larger than Li(x). We don't yet know where this happens, but it has been shown to be past 10^{20}, and is likely to be around $1.39822 \cdot 10^{316}$. This number, far greater than the number of particles in the universe (a mere 10^{75} or so), is no more than a peak in the mountainous landscape where number theorists stride. Once past their first crossing, Li(x) and π(x) exchange places infinitely many times as they draw closer and closer together.

The ratio $\frac{1}{\ln x}$ that appears in this integral turns out to have a close relative that tells us something about the distribution of twin primes—even though we still have no proof that there are infinitely many of them! A great deal of modern work allows us to say that for any number a, the number of twin primes in a run of naturals from x to x + a will be close to

$$\frac{1.3a}{(\ln x)^2}$$

(for purists, 1.3 is, somewhat more precisely, 1.3203236316. . .). This estimate predicts 584 twin primes between 10^8 and $10^8 + 150,000$, and 601 have been found. It predicts 166 between 10^{15} and $10^{15} + 150,000$, and 161 have been found.

Even the great gaps we saw yawning amid the primes can be measured by logarithms. The length of the largest prime-free gap up to the number x—call this g(x)—is well approximated by $(\ln x)^2$:

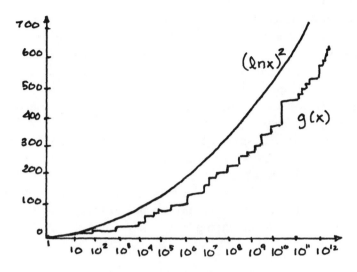

Looking at the distribution of the primes, the contemporary mathematician Don Zagier wrote that he had "the feeling of being in the presence of one of the inexplicable secrets of creation." Certainly the need to make excursions into mathematical continents so remote from the naturals in order to bring back some understanding of them, has an effect like music's on our minds: how can vibrating brass and wire, gut and air, set up such abstract poignancy within us?

What are the hints we should follow: which are beacons, which false fires? Is it important to know if the number of twin primes is indeed infinite? What about the primes that are "palindromic", like 101, 373, and 929—does it matter if there is an infinite number of them too? Or the "counting primes" such as 1,234,567,891 and 12,345,678,901,234,567,891 and (this way madness lies) 1,234,567,891,234,567,891,234,567,891? The "topping and tailing" primes: you can remove digits from either end of 739,397 and what's left remains prime; you can take as many digits as you like from the tail of 739,391,133 or from the top of 357,686,312,646,216,567,629,157 and each is still prime; do we care if there is an infinity of these? Will hidden vistas open if we one day prove the 250-year-old conjecture of Christian Goldbach, that every even number from 4 on is the sum of two primes? It is the sole memorial to the dilettante son of a Königsberg pastor, who established the pattern for the education of future Tsars, and corresponded in elegant French, German, and Latin with every renowned mathematician of his era, asking difficult questions (for it marks our kind that we know how to pose so many more questions than we know how to answer). As of 1998, Goldbach's conjecture was true up to 100,000,000,000,000—which is as nothing compared with *every* even number.

Knowing what creatures there are and what creatures there could be in this crowd of primes must surely give us a presentiment of its structure, and some corner of the pattern may be the key that will turn it into a living garden, where what is design will sort itself out from ornament.

To what, then, should we compare our present condition—as icebound as was Shackleton at the Pole? Are we like those brilliant, autistic people who understand that there must be something which facial expressions reflect and can with avid intelligence catch clues to correlate some with others, yet have no idea what the cause of such effects might be? Or more like those born blind who can yet just sense when they are facing the light but can't imagine what they will see when one day the shutters are removed? Will it be, as we hope, glorious altogether and in each part, or—with simplifying abstraction removed—speckled with unguessed dust?

∞

interlude

The Infinite and the Indefinite

I had removed the black earth's boundary stones:
Once she had been enslaved and now was free.

This is what Solon, the great Athenian lawgiver, wrote some twenty-five hundred years ago. Taking boundaries away, however, can lead from fusion to confusion and so to chaos. We know where we are when our thoughts, like our words, are sharply defined.

The Greeks had a word for the infinite and it was *apeiron* (ὄπειρον), which literally meant "without boundary" and translates equally well into "indefinite". Why should they, why should we, so concern ourselves with the endless, when it may only amount to the vague? Anaximander, who lived a hundred and fifty years before Socrates, recognized the foolishness of claiming that one element or another—earth, air, fire, water—was the source of everything else. Rather, he said, the source is the *apeiron*—as if distinction rose out of indistinction, the way it does in so many creation myths. We think this way still, seeing speciation on a grand scale evolving from the unspecified, and minutely differentiated tissues from stem cells.

The infinite disguised as the indefinite is our onlie begetter. But in this same guise it is how we imagine the world truly to be: made up ultimately not of separate objects, molecules, atoms, electrons, or quanta, but, past the ever more granular, to be as partless as the ocean, where our little prisms of selves spray up and soon enough submerge. Just as we picture continuity in the material world by rocks between boulders, stones between rocks, pebbles between stones, and sand to fill in the crevices, so we see fractions in the spaces between integers—and for fractions "ever smaller" means denominators becoming infinitely large. If the heavens are full; if everything flows; if time is a river: then not only how we began but how we go on is drenched in that ambiguous *apeiron*.

Interlude

"Tell me if ever anything was finished," da Vinci scribbled again and again over his late drawings of tumbling chaos. He tried to give some form to this chaos by representing it as cascades and waves and whirlpools, since their immensity was at least shaped by comprehensible forces. Our hope is to find some structure to the infinite, behind what might be only superficial indefinition: regularities governing infinite ensembles; powers, dominions, and thrones among its blurred degrees.

Mathematics is the art *of* the infinite because whatever it focuses on with its infinite means discloses limitless depth, structure, and extent.

four

Skipping Stones

Late in his life, Newton said: "I seem to have been only like a boy playing on the seashore, and diverting myself in now and then finding a smoother pebble or a prettier shell than ordinary, whilst the great ocean of truth lay all undiscovered before me." Although mathematics has grown exponentially since his time, we still find ourselves children standing on the edge of the limitless unexplored.

Pick up a flat stone and skim it over the water; 7, 8, ... 13 skips perhaps, before it sinks? At least we can do better than this on the ocean of numbers, following regular pulses past any horizon. Will their patterns reflect deeper truths about the working of the world? Will the ways we go about finding and confirming these patterns give insights into the ways of the mind? Consider the sequence of natural numbers, 1, 2, 3, 4, ... and their successive sums—so familiar to us by now:

$$1, 1 + 2 = 3, 1 + 2 + 3 = 6, 1 + 2 + 3 + 4 = 10, \ldots$$

$$1 + 2 + 3 + \ldots + n = \frac{n(n+1)}{2}.$$

We've proved this three times already, in ways as concrete as making patterns of dots and as abstract as induction, but have saved the most beautiful proof for last. We want to know what $1 + 2 + \ldots + n$ adds up to. Write the sum down—and then write it again, below—but this time backwards!

$$\text{Sum} = 1 + \quad 2 \quad + \quad 3 \quad + \ldots + (n-2) + (n-1) + n$$

$$\underline{\text{Sum Backward} = n + (n-1) + (n-2) + \ldots + \quad 3 \quad + \quad 2 \quad + 1}$$

And then add the columns to get:

$$(n+1) + (n+1) + (n+1) + \ldots + (n+1) + (n+1) + (n+1)$$

There are n of these terms added up—so n · (n + 1)—but of course we want only half that amount, because we've counted the sum twice; so

$$\frac{n \cdot (n+1)}{2} \ .$$

The Alcibiades impertinence of this is appealing, as also is its improvement on the head-and-tail coupling we used in the second proof (on page 30). For we need here no special case when the number of terms is odd (exceptions hint at incomplete understanding, and proof by cases at an ideal beauty not yet attained). Appealing too is the way this approach generalizes to a series of n terms beginning not with 1 but with any natural number—call it a (notice that the *second* term will be a + *1*, the *third* a + *2* and in general the *nth* term will be a + *(n – 1)*):

$$a + \quad (a+1) \quad + \quad (a+2) \quad + \ldots + \ (a+(n-1))$$
$$(a+(n-1)) + \ (a+(n-2)) \ + \ (a+(n-3)) \ + \ldots + \ a$$

which adds up to

$$2a + (n-1) + 2a + (n-1) + \ldots + 2a + (n-1)$$

or, since there are n terms,

$$n \cdot (2a + (n-1))$$

and again we want only half of this, so

$$a + (a+1) + (a+2) + \ldots + (a+(n-1)) = \frac{n \cdot (2a+(n-1))}{2} \ .$$

If, for example, a = 17 and n is 5 (the series 17 + 18 + 19 + 20 + 21), the answer should be $\frac{5 \cdot (2 \cdot 17 + 4)}{2} = \frac{5 \cdot (38)}{2} = 95$—and it is.

Why stop generalizing here? We'd like to take in every sort of sequence with our widening powers, since the mind says "all" when the eye asks "which?" What about sequences with any natural number acting as the difference, d, and not just 1? If d = 3, for example, and we start with 1, we get the sequence 1, 4, 7, 10, 13, 16 ... and the sums

$$1 \, ,$$
$$1 + \ 4 = 5 \, ,$$
$$1 + 4 + \ 7 = 12 \, ,$$
$$1 + 4 + 7 + 10 = 22 \ldots$$

Our new style of proof—leaving pictures behind and carefully maneuvering with symbols—readily accommodates this greater scope:

$$a + \quad (a+d) \quad + \quad (a+2d) \quad +\ldots+ \quad (a+(n-2)d) + (a+(n-1)d)$$

$$(a+(n-1)d) + (a+(n-2)d) + (a+(n-3)d) +\ldots+ \quad (a+d) \quad + a \,.$$

These two rows add up to

$$n \cdot (2a + (n-1)d)$$

and taking the half we want,

$$a + (a+d) + (a+2d) + \ldots + (a+(n-1)d) = \frac{n \cdot (2a+(n-1)d)}{2} \,.$$

In our example above, with a = 1 and d = 3, when n = 5 we should get

$$1 + 4 + 7 + 10 + 13 = 35$$

and in fact

$$5 \cdot \frac{(2 \cdot 1 + 4 \cdot 3)}{2} = 5 \cdot 7 = 35 \,.$$

This proof is a joy forever. Its loveliness increases not only because admiration for elegant symmetry never dies—an eternal monument to its unknown inventor—but because, like the series it describes, it ripples outward in ever new contexts. It gives us a finite grasp (in a single body and soul, as Rimbaud desired) of an infinite sequence. There are some, like the distinguished twentieth-century number theorist André Weil, for whom a conjecture once proven, like a mountain climbed, becomes no more than a trophy: another name on Don Giovanni's list. In contrast, a piece of mathematics heard as music is inexhaustibly filled with promises for the future and houses as well an inexhaustible presence, like a fugue from the *Well-Tempered Clavier*.

Sequences such as these, with a constant difference between successive terms, are called *Arithmetic Sequences*, and the sum of their terms from the first through the nth is an *Arithmetic Series*. The triangular numbers are a hybrid: a sequence of numbers, as you saw, which are successive sums in an Arithmetic Series. For the triangular numbers are 1, 3, 6, 10, . . . and so on, which are the successive sums of the sequence that starts with first term a = 1 and has difference d = 1.

What about the sequence of square numbers, which every once in a while coincides with that of the triangulars (at 1 and 36, for example)? Remember that by a clever diagonal slice we found that any square number was the sum of two consecutive triangulars. But now that the triangulars have a formal rather than visual embodiment, cleverness gives way to clockwork. For the $(n-1)^{th}$ triangular number is $\frac{(n-1)n}{2}$, and the next one, the nth, is $\frac{n(n+1)}{2}$; so multiplying each one out and adding them to each other gives:

$$\frac{n^2-n+n^2+n}{2} = \frac{2n^2}{2} = n^2 \ .$$

Is the lost glory of ingenuity the price we must pay for the gains of abstraction? Or could the interplay between the visual and the formal—the geometric and algebraic—still be fruitful? Let's follow the natural drift of our curiosity from triangular and square on to the pentagonal numbers:

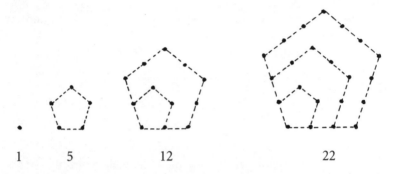

| 1 | 5 | 12 | 22 |

so this sequence begins 1, 5, 12, 22, and goes on to 35, 51, 70 . . . Aren't these just the sums we saw on page 78? But why—why should this be so? And what is the connection of the pentagonal shape to triangular numbers?

Clearly there will be no sum of two triangular numbers here, since we can't rebuild even 5 that way. But $5 = 1 + 4$; 1 is triangular and 4 is square—what if that's the breakup we're looking for? Visual ingenuity to our aid again: let's design our netted pentagons with triangles and squares in mind. This only takes some pushing in at the sides.

| 1 | $1 + 4 = 5$ | $3 + 9 = 12$ | $6 + 16 = 22$ |

That is:

$$\triangle_1 + \boxed{2} = \pentagon_2 \qquad \triangle_2 + \boxed{3} = \pentagon_3 \qquad \triangle_3 + \boxed{4} = \pentagon_4$$

and indeed

$\triangle_4 + \boxed{5} = 10 + 25 = 35$, which is $\pentagon_5$, and so, it seems, on.

Once we write out our discovery in formal terms, we can try proving it. Since it looks as if the nth pentagonal number is the $(n-1)^{th}$ triangular plus the nth square, and since the $(n-1)^{th}$ triangular is $\frac{(n-1)n}{2}$, our conjecture is:

$$\pentagon_n = \triangle_{n-1} + \boxed{n} = \frac{(n-1)n}{2} + n^2 = \frac{(n^2 - n + 2n^2)}{2} = \frac{(3n^2 - n)}{2} = \frac{n(3n-1)}{2}$$

and for n from 1 to 7, this gives us the values we want:

$$1, 5, 12, 22, 35, 51, 70 \, .$$

We have our insight, but we can't hope for a proof yet because we still need to understand exactly how any pentagonal number is built up from the previous one. Mere manipulation of letters rarely leads to seeing—but looking does.

Let's look then at how the third pentagonal number grows from the second, and the fourth from the third.

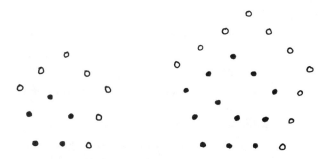

In each case, two of the old sides were extended, and three new sides fitted on to make the larger pentagon. There is one more dot per side in this new pentagon, so it looks as if we have added 3 dots per side on these 3 new sides of the third, 4 per side on the 3 new sides of the fourth. But this can't be quite right, since new sides share a dot at their corners, so we have to subtract 2 dots, giving us

new sides   dots per side   shared dots

Thus to go from the third pentagonal number to the fourth we could calculate

$$⬠_4 = ⬠_3 + 3 \cdot 4 - 2$$

$$12 + 12 - 2 = 22, \text{ as desired.}$$

In short, we must add $3n - 2$ dots to $⬠_{n-1}$ to get $⬠_n$.

We could now with relative ease prove by induction that indeed

$$⬠_n = \frac{n \cdot (3n - 1)}{2} \ ;$$

but we'll go after bigger game.

With factoring a number into its primes still ringing in our ears from the last chapter, you might by analogy wonder about reducing pentagonal numbers down to the basic triangular numbers—and of course we can. Since we know that any square number is the sum of two triangular ones, and since we saw that a pentagonal is a triangular plus a square, any pentagonal number is the sum of three triangular numbers.

Does this suggest that any hexagonal number is the sum of 4 triangulars, and so on? Wonderful Gauss made this cryptic entry in his diary on July 10, 1796:

Eureka! Num $= \Delta + \Delta + \Delta$.

Repeating Archimedes's joyous exclamation, "I've found it!", he meant that he had found a (by no means easy) proof that every natural number is the sum of at most three triangular numbers.

Can we in turn follow our triumphs thus far by coming up with a formula for the hexagonals? A kind of impatience begins to set in at this point, however, because the work promises to be strenuous—and after it we would have to begin all over again with the heptagonals, and so endlessly on. The three formulas we already have are so different from one another:

$$△_n = \frac{n(n+1)}{2}$$

$$\boxed{n} = n^2$$

$$\widehat{n} = \frac{n \cdot (3n-1)}{2} \, ,$$

that you may suspect there will be no pattern to the patterns. Mind, however, keeps clamoring for universal explanations.

Let's act like mathematicians, with faith in design and confidence in our powers to find it. Above all, let's use the art of the infinite: going, that is, for all the patterns at once. We want the general formula for what any particular figurate number is—that is, for the nth term of any polygonal sequence. The amazing thing is that this will be much easier— and infinitely less time-consuming—than doing it for each kind of polygon in turn. It is in this uncanny generalizing power that mathematics puts to shame the tailor who boasted of killing seven at a blow.

Because the formula we seek will deal with different terms of different polynomial sequences, we will need another letter to stand for *which* sort of polygon we have in mind. Let's speak of k-gons, where k can be 3, 4, 5, and so on. So a 3-gon is a triangle, a 4-gon a square, a 5-gon a pentagon. We already have a formula for the nth term of each of these, and have set our sights on a formula for the nth term of *any* k-gonal sequence. Having traded in the more colorful polygonal names for this stark way of speaking, let's make one concession more and represent the nth term of a k-gonal sequence by P_n^k.

This kind of naming and these sorts of symbols drive more people away from mathematics than teachers who tell you you're wrong because they say so. We are perfectly happy to think of someone as James Smith or even James Topaz Smith, and if his son is James Topaz Smith Junior we take that easily in stride. Should the son become a Doctor or even the Right Reverend Doctor James Topaz Smith Junior we may smile, but can handle it. Yet attach a pair of numbers to a letter and we beg for mercy. The unfamiliarity of this kind of acronym is partly to blame, as is suddenly having to read vertically—but it is the same style of naming. Like Smith, P is the family name: Polygon. k gives the branch of the family, n (like James Topaz) singles out the individual in that branch. So P_5^3 is the fifth triangular number, for example, P_2^4 is the second square number, and the seventh pentagonal number is P_7^5. What we pile on the spine of this weedy symbol will save us an enormous amount of mental energy.

Now we can indulge in the pleasures of the table once more, in hopes of insight into what is actually happening. Here are the first few entries for some k-gonal sequences:

Old Name	New Name	First	Second	Third	Fourth	Fifth
triangular	3-gonal	1	3	6	10	15
square	4-gonal	1	4	9	16	25
pentagonal	5-gonal	1	5	12	22	35
hexagonal	6-gonal	1	6	15	28	45
heptagonal	7-gonal	1	7	18	34	55

The columns are interesting but the rows even more so. Each starts with 1, and the differences between columns in the first row are 2, 3, 4, 5, ... In other words, the difference grows by 1 for each new column.

In the second row, the differences are 3, 5, 7, 9, ... Those grow by 2s. The third row's differences grow by 3s: 4, 7, 10, 13, ... and the fourth row differences—5, 9, 13, 17, ...—grow by 4s. What matters here seems to be this "growth number"; let's call it g. Put in terms of each k-gonal sequence, the differences grow by

$$1 \text{ in the 3-gonal} \quad (g = 1)$$
$$2 \text{ in the 4-gonal} \quad (g = 2)$$
$$3 \text{ in the 5-gonal} \quad (g = 3)$$
$$4 \text{ in the 6-gonal} \quad (g = 4).$$

On this scanty evidence we hazard the conjecture that in a k-gonal sequence, g will be k − 2.

It looks, then, as if we have the same hybrids in every case that we had with the triangular numbers: each term is a sum in an arithmetic series; each series starts with a = 1; the respective g is k − 2.

k-gonal	First	Second	Third	Fourth	Fifth
3-gonal	1	1+2	1+2+3	1+2+3+4	1+2+3+4+5
4-gonal	1	1+3	1+3+5	1+3+5+7	1+3+5+7+9
5-gonal	1	1+4	1+4+7	1+4+7+10	1+4+7+10+13
6-gonal	1	1+5	1+5+9	1+5+9+13	1+5+9+13+17
7-gonal	1	1+6	1+6+11	1+6+11+16	1+6+11+16+21

Look! The nth term of a polygonal sequence is the sum of the first n terms of an arithmetic sequence where a = 1 and d = k − 2.

Using the formula for the sum of arithmetic sequences which we perfected on page 78: $\frac{n(2a+(n-1)d)}{2}$, with here a = 1 and d = k − 2, we get

$$P_n^k = \frac{n(2+(n-1)(k-2))}{2}.$$

This simplifies to

$$P_n^k = \frac{n \cdot (nk - 2n - k + 4)}{2} .$$

You can check this, if you like, for some entry in our table—the fourth column, say, of the fifth row, the fourth heptagonal number, which is 34. And

$$P_4^7 = \frac{4 \cdot (7 \cdot 4 - 2 \cdot 4 - 7 + 4)}{2} = 2 \cdot (28 - 8 - 7 + 4) = 2 \cdot 17 = 34 .$$

Right as rain, and sometimes even more so.

Does this remarkable general formula turn into the particular formulas we got for triangular, square, and pentagonal numbers? With squares, for example, is the nth square number really n^2?

$$P_n^4 = \frac{n \cdot (4n - 2n - 4 + 4)}{2} = \frac{n \cdot 2n}{2} = n^2$$

and for pentagons, will we have the formula $\frac{n \cdot (3n-1)}{2}$?

$$P_n^5 = \frac{n \cdot (5n - 2n - 5 + 4)}{2} = \frac{n \cdot (3n - 1)}{2} .$$

This is a startling unity beneath such apparent diversity, which now lets you calculate in a moment the number of dots—should you care to know—in the 18th 201-gonal number, for example (321,801—a little touch of personality in the crowd).

Of course we haven't yet *proved* the conjecture we got by studying our table. Lest you think that nothing could be more boring than proving the obvious, it would be enough to remember that mathematics is the one skyscraper of thought which rises above mere opinion to utter certainty. But we can add that a proof's performance is as full of contortionists, jugglers, and high-wire acts as the world's best circus. The proof, with all its acrobatic providers, is in the on-line Annex.

Notice, here, how far we've moved from the Pythagorean tetractys into a language and style of thought where symbols (P_n^k) of symbols (3) of symbols ( ) are casual familiars. When your former self complains—as the English philosopher Thomas Hobbes did in 1656 to his contemporary John Wallis—that the page "is so covered over with the scab of symbols, that I had not the patience to examine whether it be well or ill demonstrated," your present self can answer, with Wallis: "Is it not lawful for me to write Symbols, till you can understand them? Sir, they were not written for you to read, but for them that can." Equations have come to explain more easily than sentences the structures we work among. No wonder mathematicians, like Rip van Winkle playing at bowls with the little men up in the hills, lose all sense of self and time, and on returning seem as alien as the world now seems to them.

$$\infty$$

Arithmetic sequences and series have led us along bright rays into the infinite. Even more dazzling, however, are their twins: the sequences and series called *Geometric*. A Geometric Sequence also begins with any number, but its new terms come not by adding a constant d, but through multiplying by a constant, usually called r (for "ratio").

If a = 1 and r = 2, for example, you get the larger and larger numbers

$$1, 2, 4, 8, 16, \ldots.$$

Geometric series add these all together, up to a certain term, say the 64th. Since each term is 2 raised to a power one greater than the previous one, this sum would be

$$1 + 2 + 2^2 + 2^3 + \ldots + 2^{63},$$

a finite but very large number, which anyone using the sure-fire Martingale System of betting will know from nightmares. In this system you keep doubling your bet until you win—then quit. Had you started with a dollar, you might have to go home with two—but a really bad run of luck, 64 tries long, would leave you owing more dollars than there are atoms on the earth to make them with.

When the ratio shrinks to a positive number less than one, strange and wonderful things begin to happen—especially if infinity enters again as the number of terms. Let's experiment with a = 1 and r = $\frac{1}{2}$.

$$1 + \frac{1}{2} + \frac{1}{4} + \frac{1}{8} + \frac{1}{16} + \ldots.$$

The successive terms grow rapidly smaller, their sum grows steadily larger—but will it ever become infinitely large, or as large as 19, or 3, or 2.07? In our experimental mood let's look at partial sums:

$$1 + \frac{1}{2} = \frac{3}{2}$$

$$1 + \frac{1}{2} + \frac{1}{4} = \frac{7}{4}$$

$$1 + \frac{1}{2} + \frac{1}{4} + \frac{1}{8} = \frac{15}{8}$$

$$1 + \frac{1}{2} + \frac{1}{4} + \frac{1}{8} + \frac{1}{16} = \frac{31}{16}$$

A reasonable conjecture at this point would be that the sum up to $\frac{1}{2^n}$ will be a fraction whose numerator is twice its denominator less 1: that is,

$$\frac{(2^{n+1} - 1)}{2^n}.$$

If so, the sum keeps falling just short of 2, though by less and less. This would mean that no matter how many terms we add on we will never get to 19, or 3, or 2.07—or even 2; 2 will be the reach that always just exceeds our grasp.

This is a peculiar situation, which gives us second thoughts about the infinite: an infinite number of terms whose sum is shakily finite. Does this happen only when $r = \frac{1}{2}$? Let's experiment further, taking $r = \frac{1}{3}$: 1, $\frac{1}{3}$, $\frac{1}{9}$, $\frac{1}{27}$, $\frac{1}{81}$ ... (remember: we get each new term through multiplying the previous one by $\frac{1}{3}$).

The successive sums of these are

$$1 + \frac{1}{3} = 1\frac{1}{3}$$

$$1 + \frac{1}{3} + \frac{1}{9} = 1\frac{4}{9}$$

$$1 + \frac{1}{3} + \frac{1}{9} + \frac{1}{27} = 1\frac{13}{27}$$

$$1 + \frac{1}{3} + \frac{1}{9} + \frac{1}{27} + \frac{1}{81} = 1\frac{40}{81}$$

If this is settling down to some number, as the previous series seemed to do, it is a bit more obscure—perhaps because the denominator is odd. The fraction's numerator keeps falling just short of half the denominator, as if the series were approaching $1\frac{1}{2}$ or $\frac{3}{2}$.

As with triangular numbers, we grow impatient and ask for a pattern to these patterns. Perhaps our asking is premature and the next example, with r = $\frac{1}{4}$, would help. The sequence is 1, $\frac{1}{4}$, $\frac{1}{16}$, $\frac{1}{64}$, $\frac{1}{256}$, ... and the successive sums are

$$1 + \frac{1}{4} = \frac{5}{4} = 1\frac{1}{4}$$

$$1 + \frac{1}{4} + \frac{1}{16} = 1\frac{5}{16}$$

$$1 + \frac{1}{4} + \frac{1}{16} + \frac{1}{64} = 1\frac{21}{64}$$

$$1 + \frac{1}{4} + \frac{1}{16} + \frac{1}{64} + \frac{1}{256} = 1\frac{85}{256}$$

and a shrewd guess would suggest that $1\frac{1}{3}$ was the elusive she.

Stepping back, the glimmerings of a beautiful regularity dawn. Our experimental results made 2, $\frac{3}{2}$, and $\frac{4}{3}$ the likely targets of series whose ratios were, respectively, $\frac{1}{2}$, $\frac{1}{3}$, and $\frac{1}{4}$. If we think of 2 as $\frac{2}{1}$, then it might well be that when the ratio is $\frac{1}{n}$, the sum of the *infinite* series resulting is $\frac{n}{n-1}$.

Two distinct difficulties immediately come up. How can we speak of an "infinite sum" at all, especially on the basis of what must always be finite approximations? And second, why are we even indulging in the luxury of such a conjecture on the basis of so few trials: where is the proof?

Given the two voices within us you would expect two answers (at least) to the first question. The more cautious voice says that of course those numbers we have come up with are never actually attained—but they do seem like limits drawn ever closer to—as close as you like, given sufficiently many terms. While the growing sums also get closer to numbers beyond their respective limits, these limits are the *least* such numbers approximated to from below but never reached. This voice assures us that if we say "the limit as n goes to infinity is such-and-such" we mean what we say: is on the way to but (like Chekhov's three sisters and Moscow) never gets there. Or if you like, we mean that the word "limit" abbreviates the rather complicated idea we have just expressed.

The other voice damns such caution and adds infinity to what we reckon with and on. $\frac{3}{2}$ is what *all* the terms of the first series add up to, just as language points to what it ultimately can't say. The rational numbers, which we first understood as ratios of the seemingly more concrete natural numbers, now stand revealed as embodiments of yet more fundamental, infinite processes. Remember that Brouwer saw even the naturals as belonging to a limitless, fundamental sequence unfolding through time; he also thought that the objects of the world—including other humans and even the body inhabited by the self—are no more than sequences of sequences of sensations.

These voices of what seem like angelic spectators at our human play fade out, as we come to the second question of proof. Play is, as always, the key, in its many senses: fooling around, the freedom to invent, the daring of Alcibiades and the loose play in a fitting that needs to be tightened. The stakes being played for are these: to come up with a concise way of understanding the sum of a geometric series—a way so tight that it will bear the strain of extending our understanding to sums of an infinite number of terms.

The series can start with any number a, and have as its fixed ratio any number r between 0 and 1. The successive terms will be ar, $(ar)r = ar^2$, $(ar^2)r = ar^3$, ar^4, and so on. The sum of the first n terms—let's use S for "sum" and call this S_n—would therefore be

$$S_n = a + ar + ar^2 + ar^3 + \ldots + ar^{n-1}.$$

So, for example, in the series on page 86, where $a = 1$ and $r = \frac{1}{2}$, the third term, $S_3 = \frac{7}{4}$.

The trick that worked so well before—adding to this the same sum written backwards—won't work now because we are multiplying by a constant rather than adding one. Someone fought his way through to the brilliant invention of multiplying both sides of that equation by r:

$$rS_n = ar + ar^2 + ar^3 + ar^4 + \ldots + ar^n$$

and then *subtracting* this equation from the first, after a deft realignment:

$$S_n = a + ar + ar^2 + ar^3 + \ldots + ar^{n-1}$$
$$- rS_n = \quad\quad ar + ar^2 + ar^3 + ar^4 + \ldots + ar^n$$

When subtracting we get the same effect that "canceling on the diagonal" had before, and

$$S_n - rS_n = a - ar^n$$

that is,

$$S_n \cdot (1 - r) = a \cdot (1 - r^n).$$

So

$$S_n = \frac{a \cdot (1 - r^n)}{(1 - r)}.$$

The Egyptians and Babylonians may have known this. Euclid certainly did, and gave a proof differently elegant from ours. If you find that such a vaulting proof still needs a flying buttress or two of example, you might try what the formula yields when a = 1 and r = $\frac{1}{2}$ or $\frac{1}{3}$.

Now the big question is: what will this tell us about the sum of an infinite number of terms? In our current example, will we find that the limit is 2?

With r = $\frac{1}{2}$, notice that raising r to greater and greater powers makes it steadily smaller: $(\frac{1}{2})^n$ *approaches* 0 as n *approaches* infinity. Keeping a = 1 and r = $\frac{1}{2}$, our formula is

$$S_n = \frac{1-\left(\frac{1}{2}\right)^n}{\left(1-\frac{1}{2}\right)} = \frac{1-\left(\frac{1}{2}\right)^n}{\frac{1}{2}} \; .$$

Using the word "limit" and the symbol "∞", we can express "the limit as n goes to infinity" by $\lim_{n\to\infty}$ and so can write

$$S_\infty = \lim_{n\to\infty} \frac{1-\left(\frac{1}{2}\right)^n}{\left(1-\frac{1}{2}\right)} = \frac{1-0}{\frac{1}{2}} = \frac{1}{\frac{1}{2}} = 2 \; .$$

And with r = $\frac{1}{3}$?

$$S_\infty = \lim_{n\to\infty} \frac{1-\left(\frac{1}{3}\right)^n}{\left(1-\frac{1}{3}\right)} = \frac{(1-0)}{\frac{2}{3}} = \frac{1}{\frac{2}{3}} = \frac{3}{2} \; .$$

So in general, as long as r is between 0 and 1,

$$S_\infty = \lim_{n\to\infty} \frac{1-r^n}{(1-r)} = \frac{1-0}{(1-r)} = \frac{1}{(1-r)} \; .$$

We say that our infinite geometric series *converges* to this limit. This is an astonishing victory for the finite mind over infinity.

The eye can share this triumph through a proof whose picture speaks areas, if not volumes. All you need know is that two shapes are *similar* if they have the same angles (i.e., one is a scaled-down version of the other); that if two shapes are similar, their sides are in proportion (and vice versa)— and that parallel lines meet a line crossing them at the same angle.

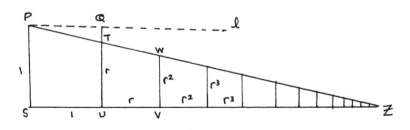

Start by making a trapezoid PSUT, with right angles at S and U (so PS and TU are parallel). Let PS = SU = 1 and UT = r.

Paste another trapezoid TUVW on its right with TU = UV = r, and the line PT continued to meet the vertical from V at W.

These trapezoids are similar (right angles at their bases, equal angles made by parallels meeting the top), so their sides are in proportion. This means that

$$\frac{VW}{r} = \frac{r}{1}$$

so VW = r².

We go on making similar trapezoids in this way—their successive right-hand sides and bases will be r³, r⁴, and so on—and the line PT will eventually intersect SU's extension at some point Z.

That base SZ is therefore the infinite sum

$$1 + r + r^2 + r^3 + r^4 + \ldots$$

Last, construct a triangle TQP, similar to $\triangle$PSZ, by drawing a line ℓ from P parallel to SZ, and extending UT to meet it at Q. PQ has length 1, and TQ = QU – TU = 1 – r.

Since $\triangle$PSZ is similar to $\triangle$TQP,

$$\frac{SZ}{PS} = \frac{PQ}{QT} .$$

That is,

$$\frac{(1+r+r^2+r^3+r^4+\ldots)}{1} = \frac{1}{1-r} .$$

There it is: seen all at once and so naturally, so convincingly, that you look back in wonder at all the sour wrangle over foundations and formal proofs. Yet such pictures as this have their critics, who would caution us to speak of them instead as "more or less proofs". Pictures can lie; at the very least they can persuade the eye to take for granted what the mind should examine in detail. How, for example, can we be sure that the line from P though R hits the line extended from SV precisely at the end of the infinite series? When the number theorist J. E. Littlewood said of a drawing that it was all the proof needed for a professional, he was suggesting that a professional would know where and how to grow the connective tissue.

Still, deductive chains descending from axioms can be so long, with so many small links so artfully forged, that we lose the sense of the whole and ache for the way a picture comprehensively, instantaneously, catches what is true, and why. Lost in the tangle of tactics we agree with Descartes that thought should move through a demonstration continuously and smoothly. Aren't we being tossed back and forth once again between the intuitions of space and time: between the visual cortex and powers we read as greater precisely because they work with the unembodied?

The different styles of proof you have just seen are more on a par with one another than were those in Chapter One, where the visual had it all over its algebraic equivalents. Is that because we are now more experienced with the algebraic, or because our standards of proving have grown higher—or is it just a matter of equal stimulation to different centers of pleasure in the mind? And could a visual proof still trump an algebraic one?

Consider a maverick infinite sequence that is neither arithmetic nor geometric:

$$\frac{1}{2} + \frac{2}{4} + \frac{3}{8} + \frac{4}{16} + \frac{5}{32} + \ldots$$

where each term is of the form $\frac{n}{2^n}$. What is *its* sum? If you picture this as did Nicole d'Oresme, Bishop of Lisieux, around 1350, the answer suddenly stares disconcertingly back at you: the left-hand tower shows the sequence vertically; the second shows the sum of each horizontal row in turn—the third brings the second dramatically down to earth.

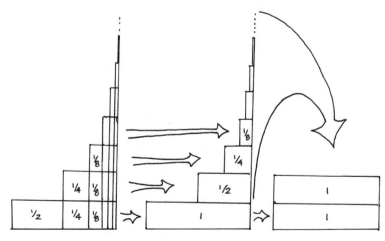

The sum is 2. What is not a little disturbing is that read from right to left, a *finite* area is extended *infinitely* in space (those endlessly rising blocks). When three centuries later Evangelista Torricelli—the inventor of the barometer—took this paradox one dimension higher by depicting an infinite solid whose volume was in fact finite (the curve $\frac{1}{x}$ spun around its axis, from 1 to infinity),

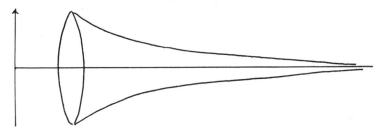

Thomas Hobbes wrote: "To understand this for sense, it is not required that a man should be a geometrician or logician, but that he should be mad." To make Hobbes's outrage more vivid, realize that a finite amount of paint poured in would coat its infinite surface (a paradox for mathematicians only; physicists know that molecules of even the finest oil will seep just so far down the trumpet's diminishing diameter).

Going back to our geometric series, why should we have insisted that r's value lie between 0 and 1? Certainly in our visual proof only that would make the line PT meet the base at Z. The choice of r, however, didn't trouble George Peacock, of the Principle of Permanence. It must have been one day before breakfast when he reasoned that if r = 1, the left-hand side of

$$\frac{1}{1-r} = 1 + r + r^2 + r^3 + r^4 + \dots$$

became $\frac{1}{1-1} = \frac{1}{0}$, which he was happy to call ∞, while the right-hand side would be $1 + 1 + 1 + 1 + \dots$ forever, which is infinite indeed. And if r were 2, for example, $\frac{1}{1-r}$ would be –1, and as for the right-hand side. . . the right-hand side would be $1 + 2 + 4 + 8 + \dots$, which he complacently described as *more* than infinity. The third impossible thing he did that morning was to accept this equality of –1 with "more than infinity", and the fourth was to say that therefore the equality sign had a meaning beyond mere numerical identity. He went on to urge us to accept his reasoning in order to avoid an embarrassing multiplicity of cases—and sixth, he said that rejecting his point of view would deprive almost all algebraic operations of their certainty and simplicity.

Peacock was writing at a time when astonishing revelations kept tumbling out of the study of sequences and series, like harvest from an upended cornucopia. Since mathematics is freedom, why shouldn't he have felt that whatever isn't expressly forbidden is allowed? But from that cornucopia had also fallen the opened Pandora's box of divergent series: series which, as you added up their terms, failed to converge toward any limit but grew uncontrollably large; and which seemed to come wreathed with the legend: "Everything is forbidden that isn't expressly allowed."

Geometrical series with r greater than 1 were a single example among many. Could there be series whose successive terms grew smaller and smaller, yet whose sum grew ever larger? The same common sense that tells us the earth is flat also tells us this couldn't be. Take for example the sequence $\frac{1}{n}$, as n runs through the natural numbers:

$$1, \frac{1}{2}, \frac{1}{3}, \frac{1}{4}, \frac{1}{5}, \ldots$$

If you look at the growing sum, called the *Harmonic Series*—

$$1 + \frac{1}{2} + \frac{1}{3} + \frac{1}{4} + \frac{1}{5} + \ldots$$

it may very much remind you of our familiar

$$1 + \frac{1}{2} + \frac{1}{4} + \frac{1}{8} + \frac{1}{16} + \ldots$$

whose limit is 2. What is the limit of this one? More than 2, since the first four terms add up to $2\frac{1}{12}$, but perhaps not very much larger: the first ten terms—from 1 to $\frac{1}{10}$—give a total slightly less than 3, and every new weight tips this balance less and less.

Freedom in mathematics is like freedom everywhere: under law. We enter into the covenants expressed in our axioms in order to protect our freedom. The perplexity is that as we explore and develop new territories, we don't quite know what those laws are—for while we carry our axioms into the wilderness with us, they may not contain charms to subdue the strange creatures we meet. So with infinite series. It took a combination of daring and nostalgia to master this sum of the terms $\frac{1}{n}$: nostalgia in the instinct to compare it with what we know and daring in the willingness to do so without reservations. And then, that Alcibiades touch of ingenuity, to find among the familiar forms just those that would give shape to this Proteus.

$1 + \frac{1}{2}$—part of our past. $\frac{1}{3}$—had the next term been $\frac{1}{4}$ we would have been on home ground. Well, $\frac{1}{3}$ is greater than $\frac{1}{4}$, so $1 + \frac{1}{2} + \frac{1}{3}$ is just a touch greater than $1 + \frac{1}{2} + \frac{1}{4}$. Ah—and $1 + \frac{1}{2} + \frac{1}{3} + \frac{1}{4}$ is precisely that touch greater than $1 + \frac{1}{2} + \frac{1}{4} + \frac{1}{4}$. But $\frac{1}{4} + \frac{1}{4}$ is $\frac{1}{2}$ again: and that was the key to something uncanny which Nicole d'Oresme discovered. The first four terms of our new series add up to more than $1 + \frac{1}{2} + \frac{1}{2}$. Thinking in terms of successive halves, the next *four* terms are each greater than or equal to $\frac{1}{8}$, so their sum contributes more than $\frac{1}{2}$ to the total; and the next *eight* (each being greater than or equal to $\frac{1}{16}$) contribute more than $\frac{1}{2}$ again.

$$1 + \frac{1}{2} + \underbrace{\frac{1}{3} + \frac{1}{4}}_{> \frac{1}{2}} + \underbrace{\frac{1}{5} + \frac{1}{6} + \frac{1}{7} + \frac{1}{8}}_{> \frac{1}{2}} + \dots$$

compare each term with $\frac{1}{4}$ compare each term with $\frac{1}{8}$

The next run of 16 terms will add its own total of more than a half, as will the subsequent 32 terms. So this series must slowly edge its way up and past any sum of halves, and hence past any number whatever: it is unbounded and must diverge!

A wholly new set of instincts had to be developed now to cope with these innocent-seeming infinite series. Which were enemies and which were friends, and unto whom? Subtle and super-subtle tests were devised

to sniff out the series that converged; and what they converged to; and how much information could be extracted from divergent series. You prayed that the series you were exploring would turn out to be convergent. There is a story like a Biedermeier painting of the famous Hermann Minkowski, walking through the streets of Göttingen, Hilbert's Mecca for mathematicians, in the early years of the twentieth century. On Weenderstrasse he saw a student he didn't know, deep in thought. Minkowski went up to him, patted him on the back and said: "It is sure to converge."

Hermann Minkowski (1864–1909)

Wonders appeared in these woods. Important numbers like π emerged from the caterpillar of an infinite series:

$$\pi = 4(1 - \frac{1}{3} + \frac{1}{5} - \frac{1}{7} + \frac{1}{9} - \frac{1}{11} + \dots)$$

and e—the base of the natural logarithms which we met in the last chapter—is an irrational which is approximately

2.718281828459045 . . .

but is *precisely* the sum, as n goes from 1 to infinity, of $\frac{1}{n!}$:

$$\frac{1}{1!} + \frac{1}{2!} + \frac{1}{3!} + \dots$$

or, as it is more concisely written:

$$e = \overset{\infty}{\underset{n=1}{S}} \frac{1}{n!}$$

Yet there were creatures with talons here too; for while Goya was right in saying that monsters arose when reason slept, Poincaré remarked in 1899 that "logic sometimes makes monsters." He had in mind bizarre functions that fluttered around awkward series. "The divergent series are the work of the devil," the Norwegian mathematician Niels Abel had written more than half a century before; ". . . these series have produced so many fallacies and paradoxes. . ."

It took more than a century to domesticate these grotesques. Even some series that properly converge can have rowdy children. Here is a telling example, uncomfortably close to those we have scraped an acquaintance with. Instead of the harmonic series

$$1 + \frac{1}{2} + \frac{1}{3} + \frac{1}{4} + \frac{1}{5} + \dots ,$$

which we now know diverges, let's look at its cousin, where every term with an even denominator is negative:

$$1 - \frac{1}{2} + \frac{1}{3} - \frac{1}{4} + \frac{1}{5} - \frac{1}{6} + \dots$$

this series actually *does* converge: the successive sums keep hopping right and left, in ever-diminishing steps, around the limit they approach.* Let's call that limit x, so that

$$x = 1 - \frac{1}{2} + \frac{1}{3} - \frac{1}{4} + \frac{1}{5} - \frac{1}{6} + \dots$$

Now using our Commutative Axiom for addition (A2) again and again, rewrite the series with some terms interchanged as follows:

$$x = 1 - \frac{1}{2} - \frac{1}{4} + \frac{1}{3} - \frac{1}{6} - \frac{1}{8} + \frac{1}{5} - \frac{1}{10} - \frac{1}{12} + \frac{1}{7} - \dots ;$$

in other words, two negative terms in a row after each positive one.

---

*This idea of closing in on the limit from both sides suggests that in our geometric series we could let r be negative, as long as it is greater than −1: a geometric series with ratio r converges as long as −1 < r < 1.

We can now group the terms (using the Associative Axiom A1) to give us

$$x = \left(\frac{1 - \frac{1}{2}}{2}\right) - \frac{1}{4} + \left(\frac{1}{3} - \frac{1}{6}\right) - \frac{1}{8} + \left(\frac{1}{5} - \frac{1}{10}\right) - \frac{1}{12} + \left(\frac{1}{7} - \frac{1}{14}\right) \ldots$$

but $(1 - \frac{1}{2}) = \frac{1}{2}$, $(\frac{1}{3} - \frac{1}{6}) = \frac{1}{6}$ and so on, so we get:

$$x = \frac{1}{2} - \frac{1}{4} + \frac{1}{6} - \frac{1}{8} + \frac{1}{10} - \frac{1}{12} + \frac{1}{14} \ldots$$

we can factor $\frac{1}{2}$ out of each term of the right-hand side, so that

$$x = \frac{1}{2}\left(1 - \frac{1}{2} + \frac{1}{3} - \frac{1}{4} + \frac{1}{5} - \frac{1}{6} + \ldots\right)$$

and then notice that what is in parentheses is the very series we began with, x. Hence

$$x = \left(\frac{1}{2}\right)x \, .$$

The only number that satisfies this equation is x = 0.

But x couldn't *possibly* be 0! The first two terms, $1 - \frac{1}{2}$, give us $\frac{1}{2}$, and successive pairs of terms $(\frac{1}{3} - \frac{1}{4})$, $(\frac{1}{5} - \frac{1}{6})$ only add more positive values to it. We seem to have to conclude that 0 is somewhat bigger than $\frac{1}{2}$; easy for Peacock, perhaps, but not for us. It took some time to understand that our axioms for addition, which are always defined with combinations of two or three terms, may not—*do* not—extend to an infinite number of terms all at once. Here again, "an infinite number" is totally different from "a great many". Such a series as this can in fact be so rearranged as to make it converge to *any number you choose*. We need to accord with the character of new terrain as we edge our old ways into it, just as those man-made rectangular plots you see in flying over the American west yield to the givens of mountain and desert.

∞

Two series which look very much alike have behaved extremely differently:

$$1 + \frac{1}{2} + \frac{1}{4} + \frac{1}{8} + \frac{1}{16} + \ldots$$

converges to 2.

$$1 + \frac{1}{2} + \frac{1}{3} + \frac{1}{4} + \frac{1}{5} + \dots$$

diverges.

What if, before we leave this forest, we look at one of the most fabulous beings in it—a series akin to both of these:

$$\frac{1}{2} + \frac{1}{3} + \frac{1}{5} + \frac{1}{7} + \frac{1}{11} + \dots,$$

the series whose terms are $\frac{1}{p}$, for each prime p in succession. Does it converge or diverge? How can we tell, when we have so feeble a grasp of the succession of primes? The fact that each of its terms is less than the corresponding terms of the divergent series $1 + \frac{1}{2} + \frac{1}{3} + \frac{1}{4} + \frac{1}{5} + \dots$ gives hope that this one might converge. The fact that its terms begin, at least, equal to the second and then larger than the third, fourth ... terms of the convergent series $1 + \frac{1}{2} + \frac{1}{4} + \frac{1}{8} + \dots$ somewhat damps this hope. It is out there in no-man's-land, and something immensely crafty would have to be done in order to wheedle its secret from it.

It was Euler who first gave a proof of its fate, and in the twentieth century another, wonderful proof was put together from spare parts by Paul Erdos—the man who spoke of The Book with the most beautiful proofs in it. You will find in the Appendix a third, dashing proof and can judge for yourself whether it belongs in The Book. The steep ascents here and there in it will give you a good sense of what makes a proof difficult; of why the difficulties are worth it; and of how piecemeal engineering can be elevated to an art.

We will end this chapter with a smoother stone than ordinary, and skipped in a different direction from those whose flight we have been following. For here our sequence will take the form of infinitely towering exponents—the image of mathematics. It begins modestly enough:

$$x^x$$

then $(x^x)^x$, and $((x^x)^x)^x$, and so up and up forever:

We innocently ask: if this proud tower were equal to 2,

$$= 2,$$

what would x have to be?

The most reasonable response to this question is to throw up your hands in despair. It reverberates like the knocking at the gate in *Macbeth*: danger and darkness without. At a second glance, however, you might think: well, x at least can't be 1, since $((1^1)^1)^1$ and so on forever is just 1, and hence is too small. Nor could x = 2, since $2^2 = 4$, and $4^2 = 16$, galloping from the very first past the 2 this tower is supposed to equal. So if x is anything, it lies somewhere between 1 and 2. But then the waiting darkness closes in, as with the series of reciprocals of primes, $\frac{1}{2} + \frac{1}{3} + \frac{1}{5} + \frac{1}{7} + \dots$. Perhaps there simply is no answer. Or—heeding Hilbert—there may be one, but no human will ever have the wit to find it. Yet listen to Hilbert fully: there is no problem that cannot be solved. It may take art, ingenuity, insight to solve this one, but solve it we shall.

It may only take looking—and looking from an unusual angle. That whole left side—that endless tower of x's—is 2. Endless tower... but it is just as endless if we begin at the second floor as at the first: in fact, the tower *without* its base (the first x) is identical to the tower with it. This is a time, then, when the strange ways of the infinite come to our aid: that equality wouldn't appear with any finite number of x's, no matter how large. This means that all those compiled x's from the second on, being exactly the same as the whole tower (so strange are the ways of the infinite), *must likewise equal 2* (since that's what the whole tower is equal to). Making this substitution, we have:

$$x^2 = 2.$$

So x would turn out to be the wild presence that haunted our first chapter: $\sqrt{2}$. And, indeed, when x = $\sqrt{2}$, Minkowski's reassurance won't be out of place: the sequence

$$\sqrt{2}^{\sqrt{2}^{\sqrt{2}^{\sqrt{2}^{\sqrt{2}^{\sqrt{2}^{\sqrt{2}^{\sqrt{2}}}}}}}}$$

will converge.

five

Euclid Alone

When the poet Robert Graves was a student at Oxford, his tutor was the eminent classicist Gilbert Murray. "Once," wrote Graves,

> as I sat talking to him in his study about Aristotle's *Poetics*, while he walked up and down, I suddenly asked: "Exactly what is the principle of that walk of yours? Are you trying to avoid the flowers on the rug or are you trying to keep to the squares?" He wheeled around sharply: "You're the first person who has caught me out," he said. "No, it's not the flowers or the squares; it's a habit I have got into of doing things in sevens. I take seven steps, you see, then I change direction and go another seven steps, then I turn around. I consulted Browne, the Professor of Psychology, about it the other day, but he assured me that it isn't a dangerous habit. He said: 'When you find yourself getting into multiples of seven, come to me again.'"

Since we seem ourselves to have gotten from numbers into sequences of numbers, perhaps it is time to take Browne's advice and switch to the squares and flowers in the carpet: the delicate patterns of Euclidean plane geometry. The touches you have seen already hint at its power and sweep, its combination—like Greek architecture—of ingenuity and formality, with a sense of proportion subduing matter to design.

What never emerged in those touches is a stunning peculiarity. In order to reach conclusions about very finite figures, very near at hand, Euclid has to make an assumption involving the infinite. The fifth of his neat set of postulates says in effect that if ℓ is a line and P is a point not on it, then there will be one—and only one—line through P (call it m) which is parallel to ℓ.

Parallel: that is, m and ℓ will never intersect. Nowhere, through all the infinite extent of the plane, will you ever come on a point common to both.

This postulate made the Greeks uneasy. Expert seamen though they were, their longest voyages always turned round; their longest epics might take heroes to the Hesperides or the Phaeacians, but these were only a sleep away. To invoke the infinite was to call up Formlessness and the Void, to detach mind from experience. Yet there was no way around it: you couldn't prove as homely a truth as this, that the angles in a triangle added up to a straight angle (or as we would say, to 180°), without the parallel postulate to add divine strength to your mortal arm.

For if you picture triangle ABC with its base, BC, lying on line ℓ:

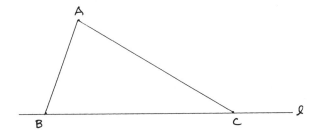

and label its interior angles a, b, and c:

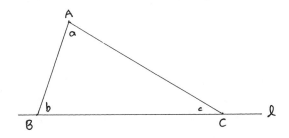

then using the parallel postulate you can draw the one and only line m through A parallel to ℓ (in symbols, m ∥ ℓ):

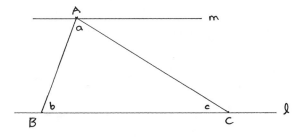

This creates two new angles—call them d and e—flanking a:

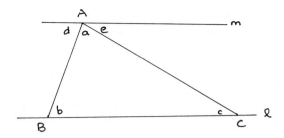

and d, a, and e evidently add up to a straight angle. Now for a touch of human devising. Extend the line making side BA upward—call this line n:

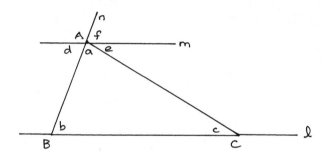

then by what we claimed on page 90—that parallel lines meet a line crossing them at the same angle—the angle we have called f, between lines n and m, must be the same as the angle b between lines n and ℓ:

$$\angle f = \angle b .$$

But when two lines, such as n and m, intersect at a point like A,

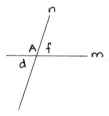

the "opposite angles" f and d are the same:

$$\angle d = \angle f ,$$

so

$$\angle d = \angle f = \angle b ,$$

that is,

$$\angle\, d = \angle\, b\,.$$

In just the same way—extending side CA upward—form $\angle\, g$:

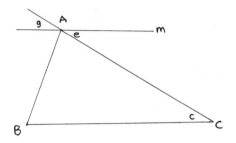

$$\angle\, g = \angle\, c, \text{ and } \angle\, e = \angle\, g\,,$$

so

$$\angle\, e = \angle\, g = \angle\, c\,,$$

that is,

$$\angle\, e = \angle\, c\,.$$

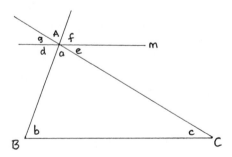

Since $\angle\, d + \angle\, a + \angle\, e$ is a straight angle, so is $\angle\, b + \angle\, a + \angle\, c$: the sum of the angles in a triangle (as the Pythagoreans may have been the first to prove) is a straight angle (180°).

While you had to accept the parallel postulate as true (the caste mark of postulates), there was no reason for you to believe us here, or on page 90, when we said that parallel lines meet a crossing line at the same angle; or that opposite angles are equal. You could appeal to your intuition; or you could work your way easily back from this theorem in Euclid to those earlier ones, following in the steps of Thomas Hobbes (whom you now meet for the third time):

He was forty years old before he looked on Geometry; which happened accidentally. Being in a gentleman's library..., Euclid's *Elements* lay open, and 'twas the 47th *El.* libri I [the Pythagorean Theorem]. He read the proposition. "By God!" said he (he would now and then swear by way of emphasis), "this is impossible!" So he reads the demonstration of it, which referred him back to such a proposition; which proposition he read. That referred him back to another, which he also read. *Et sic deinceps* [and so in order] that at last he was demonstratively convinced of that truth—this made him in love with geometry.... I have heard Mr. Hobbes say that he was wont to draw lines on his thigh and on the sheets, abed, and also multiply and divide.

Parallel lines give the Euclidean plane its character. Intersecting lines, like those netted around our plain triangle, are tactical thought made visible. And three lines setting off each on its separate mission, yet happening to concur at a single point, mark rare occasions: beams from a beacon signaling a significant event. So too any pair of points lie on a line (another postulate); but three differently defined points that happen to be collinear are the sign of deeper processes at work.

The beauty of Euclid's approach lies in building up his geometry from the simplest polygon there is: the triangle, that closed laboratory cut out of the infinite plane. He begins by laying down when two triangles, however differently situated, are the same: "congruent", in the patois of the trade, written ≅. This means that their corresponding parts—side-lengths and angle-measures—are equal, so that you could, if you wanted, fit one on top of the other and see only a single copy.

Instead of having to check, every time, each of the three pairs of sides and each of the three pairs of angles, Euclid sets down as a postulate (is it self-evident?) that if just two pairs of corresponding sides and the angles between them are equal, then the rest of the pairs must be equal too: the triangles are congruent.

Here AB = DE, BC = EF, and ∠ B = ∠ E :

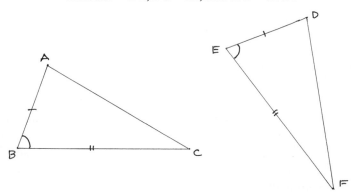

So by this side-angle-side (SAS) postulate, ∆ABC ≅ ∆DEF. From this he is able to deduce that angle-side-angle (ASA) will also be enough to guarantee congruence:

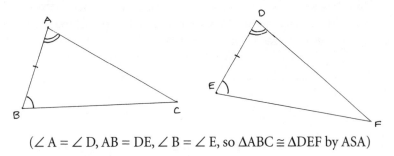

(∠ A = ∠ D, AB = DE, ∠ B = ∠ E, so ∆ABC ≅ ∆DEF by ASA)

This deduction is needed because some combinations don't suffice, such as SSA: two pairs of congruent sides and a pair of congruent angles *not* lodged between them—because, as you see below, those conditions allow you to create two *noncongruent* triangles: ABC and ABC′.

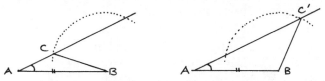

In the special case of right triangles, the equality of one pair of legs, and of the respective hypotenuses, suffices:

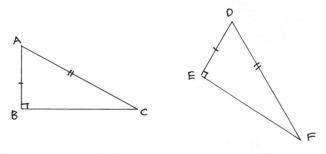

(AB = DE, AC = DF, so right ∆ABC ≅ right ∆DEF)

Do these symbols and abbreviations help or hinder? They are meant to make language transparent so that the ideas will shine through—but at first they may act like a ratchet, catching at thought. As proofs lengthen from a few to many steps, we trust more and more to the notation to carry our concentration forward. As with written music, chess manuals, or the shorthand of a trade, we come with practice to take in ever larger sweeps at a glance. The aim is always to aid intuition, not to fossilize insight into formalism.

The letters, markings, angle signs, and congruence signs belong to the proving, not to the triangles themselves. What have they in their pockets save their angle sum? With triangles, what you see is what you get. They may be embodied in a corner brace down in the basement, or in three stars a million light-years away—but this atom of plane geometry is as innocent of secrets as a baby's face.

Of course, the faces of babies no longer seem quite as innocent as they did in our pre-lapsarian youth, since their minute features must develop as the genetic code threaded through them dictates—so not even such metaphors can come close, it seems, to the emptiness of a triangle. Its sides may lengthen or shrink, its angles narrow or widen, but these infinite variations on its simplicity serve only to emphasize how thoroughly we know it, inside and out.

Let's just tickle this emptiness a bit before moving on, to see if virtual particles pop into its empty space. If you find the midpoint D of one of a triangle's sides, such as AB, and set up (see the Annex) a perpendicular to AB there—call it ℓ—(in symbols, $\ell \perp AB$),

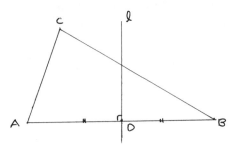

then any point Q on ℓ will be as far from A as it is from B, and conversely any point equally far from them will lie on ℓ. You can get a feel for why this is so if you think of ℓ as a flagpole and lines from Q to A and B as guy wires, holding it steady. If you prefer a formal proof, Euclid will oblige.

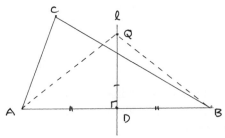

In this diagram (it looks like a specific triangle, but stands, as in a morality play, for All), ΔADQ and ΔBDQ are congruent by SAS, because AD = BD (since D is the midpoint of AB), QD is equal to itself, and the angle trapped between these corresponding sides is in each case a right angle (since that is what it means for one line to be perpendicular to another). Hence AQ = BQ. Try proving the converse yourself.

Notice that nothing depended on a specific length for QD, so Q, as it slides up and down ℓ, always stays as far from A as from B. Well, what of it? This is just artifice layered on empty form. True. But since there was nothing special about the side AB, the same must hold for the perpendicular bisector m of the side AC, erected at its midpoint E. ℓ and m can't be parallel (if they were, CAB would be a straight line, i.e., ∠ A would be a straight angle, which would blow our triangle apart), so they must meet at a point O:

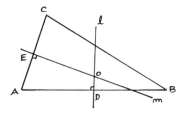

Once again, the same must be true for n, the perpendicular bisector of BC, erected at its midpoint F. It will meet ℓ at some point R and m at some point S,

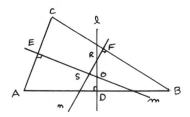

making a new little triangle ORS—but does such a triangle ORS really exist? O is on ℓ, so it is equally far from A and B. But O is also on m, so it is equally far from A and C:

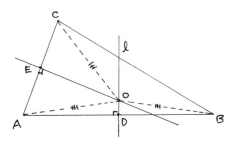

This can only mean that O is equally far from B and C—so O must *also* be on the perpendicular bisector of BC! This *transitivity* of equality, nothing more, shows us that the triangle's three perpendicular bisectors are concurrent: one of those significant events. Any triangle of necessity carries invisibly around with it a specific point that is equidistant from its three vertices. This point may lie inside the triangle, as in our diagrams, or outside it, when the triangle is obtuse (i.e., has an angle greater than a right angle, 90°):

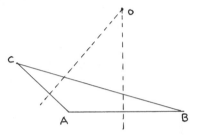

As you might begin to suspect, in the third case—when one of the angles is a right angle—this point O lies on a side: on the hypotenuse. This important fact will play a key role later—its proof is in the Appendix.

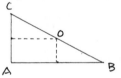

Let a skeptical friend scatter three non-collinear points A, B, and C as he chooses; you can always astound him by drawing an elegant circle through them. Join those points by straight line segments, making a triangle; erect the perpendicular bisectors of any two of these sides—and where they meet at O will be the center of the circle you seek, whose radius will be the length from O to any vertex, such as OA:

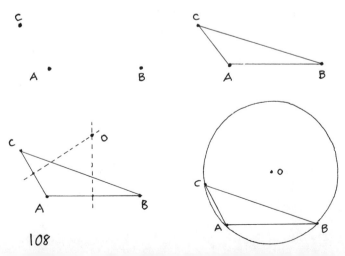

This circle is called the triangle's *circumcircle*, since it is circumscribed about it; and O is therefore called the *circumcenter*. Wipe that tabula rasa off the triangle's face: it now comes equipped with its circumcenter, like Orion with the Dog Star:

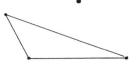

Has a triangle other dark stars just waiting to be made visible? Since its only features are sides and angles and we've just looked at the side-bisectors, let's see how the angle-bisectors behave—perhaps they too concur.

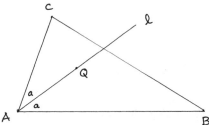

If ℓ is the bisector of the angle at A, any point Q on it will be equally far from the two sides AB and AC (a nice counterpoint to the side-bisectors). As before, let's give our intuition a formal basis. The distance from a point to a line is the perpendicular to that line from the point, so the distance from Q to AB is the length QD (since QD ⊥ AB) and from Q to AC it is QE (QE ⊥ AC), where D and E are the feet of their respective perpendiculars:

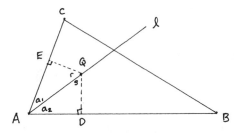

We want to show that QD = QE, and the easiest way to do this is to make them corresponding parts of congruent triangles. In this situation, complementary to the first, we'll use the complementary congruence technique of ASA. $\angle a_1 = \angle a_2$ in $\triangle$AQD and $\triangle$AQE, and certainly AQ = AQ. If we could just show that $\angle r = \angle s$ ... But the right angles are equal, and the sum of the angles in each triangle is 180°, so

$$180° - (\angle a_1 + \text{right angle}) = 180° - (\angle a_2 + \text{right angle}),$$

that is,

$$\angle\, r = \angle\, s.$$

The two triangles are congruent by ASA, so their corresponding parts are equal—among them, QD = QE. A point on the angle bisector is equally far from the sides of the angle it bisects.

The bisector of ∠ B will meet ℓ at some point I (as before, were they parallel the triangle would, impossibly, have more than 180° in it):

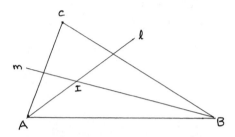

With our newly acquired sophistication, let's draw CI and hope it too is an angle bisector—hope that any point on it is equally far from CA and CB:

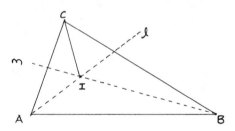

We'll think transitively, as before, and see where it leads us.

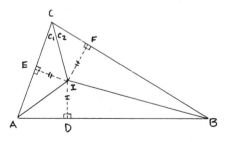

Because I is on the bisector of ∠ A, it is equally far from AB and AC: ID = IE. Because it is on the bisector of ∠ B, it is equally far from BA and BC: ID = IF. So IE = IF. We want $\Delta CIE \cong \Delta CIF$. We have a right angle in each, IE = IF and the hypotenuse IC is equal to itself—so by the "hypotenuse-leg" theorem, $\triangle$CIE $\cong$ $\triangle$CIF.

This means their corresponding parts are equal—and among these equal pairs, $\angle c_1 = \angle c_2$. Hence CI is indeed the bisector of $\angle$ C. Once again, three lines with special functions are concurrent, and the point I where they concur is called the triangle's *incenter*, because with I as center and ID, for example, as radius, you can draw the *incircle*, fitting snugly inside the triangle, whose sides will just touch (be *tangent* to) it.

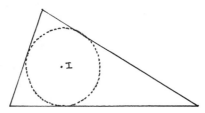

How minuet-like these reciprocal movements have been: side-bisectors, the circumcircle and its circumcenter O; angle-bisectors, the incircle and its incenter I. Remove the overlay of proof and what remains are the triangle's secret sharers.

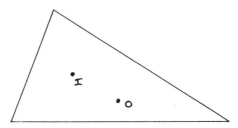

Are there more stowaways under the decks? You might expect them to be harder and harder to roust out. Well, what lines must accompany a triangle? The line from a vertex to the midpoint of the opposite side, for example, called a *median*. Here is the median ℓ from A to the midpoint D of BC.

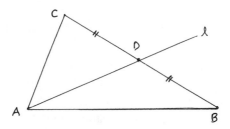

Notice that the median is a new sort of line: it certainly isn't, in general, the perpendicular bisector of side BC, nor the bisector of $\angle$ A (though in the special case of an equilateral triangle it will be both). The median, m, from C to the midpoint, E, of AB will meet ℓ at some point—call it G:

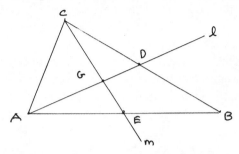

We begin to suspect that the median from B to F, the midpoint of AC, will pass through G, though there seems no immediate reason why it should.

For the sake of our intuition, let's do something Euclid would never have done and imagine our triangle actually cut out of a thin sheet of metal, with its mass spread out uniformly; and then picture balancing this triangular gusset on a knife-edge.

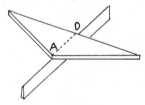

It just feels right that the knife-edge will run from a vertex, such a A, to the midpoint D of the opposite side—in other words, will be the embodiment of a median—because that way the gusset's mass will be equally divided.

This would be true if we ran our knife-edge from B to the midpoint F of the opposite side—so these two knife-edges will intersect at a point G.

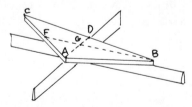

G is the triangle's *centroid,* or center of gravity: you could spin the triangle around on a pinpoint put under G; if you hung it from a thread fastened at G it would lie level*, which means that the median knife-edge from C on which the triangle balances must also pass through G.

---

*For a proof that any triangle has a centroid, through which all such mass-balancing knife-edges must pass, see the Appendix.

Metal and gussets and knife-edges don't belong to mathematics—
nor does this "proof". It was meant only to strengthen belief, not yield
certainty, as skirling pipes collect our powers for the battle ahead. Yet
while we are in the mode of physical analogy, let's press it further to see
whether it can tell us just where this centroid is.

Unequal masses won't balance at *equal* distances from a seesaw's fulcrum, but the "law of the lever" says they will balance when the distances
are adjusted so that the product of one mass times its distance from the
fulcrum equals the product of the second mass times *its* distance.

$$m_1 \cdot d_1 = m_2 \cdot d_2$$

$$2 \times 6 = 3 \times 4$$

Keeping this in mind, let's go back to our solid triangle and heat it so
much that the metal becomes molten, and then draw off the mass equally
to the three vertices. To keep the triangle's shape, imagine thin wires
connecting the three blobs at A, B, and C, which have cooled into beads
that can slide on these wires:

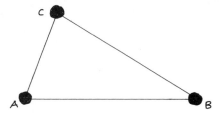

Move the blobs B and C to the midpoint D of the wire between them,
and solder in a wire from A to D.

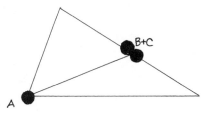

The centroid G is somewhere on this new wire. Where? D now has twice
the mass of A; so, by the law of the lever, the balance-point between D

and A must be twice as far from A as from D: in other words, the centroid G is two-thirds of the distance from A to D. This will be true for any median: the centroid lies $\frac{2}{3}$ of the distance from the vertex to the midpoint of the opposite side.

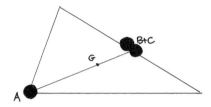

At this point you may say: no need to go on—we have our proof and a nifty bit of thinking it was too! All of a sudden the enormous gap between intuition and formal proof opens again—more vividly than ever. On the one hand, you can feel the weight of conviction almost as palpably as you can feel the weight of those metal beads. How could a triangle *not* have a centroid, and how could it not be just where we found it? On the other, temperate voices remind you that if visual proofs need interpretation, physical ones need even more; that the "law of the lever" doesn't precede but follows from mathematics; that we have let too many assumptions go unchallenged here (that mass can be replaced by masses concentrated at points; that mass tells us about area, and area about location of lines). Form itself lies behind shaped matter, and mathematics concerns itself with the play of form.

Like Archimedes, then—who looked to physics for his insights but to mathematics for his proofs—let's carry our insight back into geometry and find a proof that a triangle's medians are concurrent. We need only borrow from Euclid two early results: (1) in a triangle, the line joining the midpoints of two sides is parallel to, and half the length of, the third side;

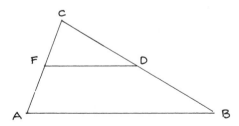

(here the line FD is parallel to AB, and half its length)

and (2) in that interesting shape, a parallelogram (a four-sided figure with one pair of sides parallel and equal, or—what turns out to be the same thing—the sides parallel in pairs), the diagonals bisect each other:

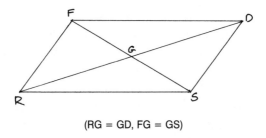

(RG = GD, FG = GS)

Confident of the outcome, we now begin. In △ABC, with D and F the midpoints of BC and AC, respectively, draw DF and the medians AD and BF, intersecting at G.

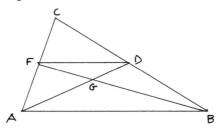

Let R be the midpoint of AG, S the midpoint of BG, and draw FR, RS, and SD.

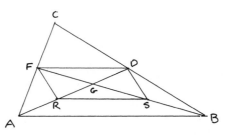

Now we'll make double use of result (1): in △ABC, DF ‖ AB and DF = $\frac{1}{2}$ AB; and in △AGB, RS ‖ AB and RS = $\frac{1}{2}$ AB. By all-powerful transitivity, DF ‖ RS and DF = RS, so RSDF is a parallelogram.

We know from result (2) that its diagonals bisect each other, so RG = GD. But R was the midpoint of AG, so in fact AR = RG = GD; that is, G is $\frac{2}{3}$ of the way from A to D. If we repeat this construction with medians AD and CE,

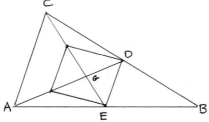

we will get exactly the same result, with the diagonals intersecting $\frac{2}{3}$ of the way from A on AD. Since there is only one point on AD which is $\frac{2}{3}$ of the distance from A, this point is again G: which means the median CE passes through G, and we have shown not only that the medians are concurrent but *where* they concur.

Part of the beauty of this proof lies in making such potent use of such a simple fact as that a line-segment has only one point on it which is two-thirds of the way from one of its ends; another part lies in how it leans on—but then straightens up from—an intuition derived from physics.

So rich is mathematics that more—and more various—proofs grow in it than ways of making your point in rolling dice or devices for emerging from the middle game in chess. This means that taste, personality, and cast of thought can be accommodated. The proof you've just seen suits lovers of symmetry; should you, however, have been seduced by infinite sequences, a custom-tailored proof is in the on-line Annex.

However you choose to prove it, another star winks on in the night sky. Our triangular Orion now, we see, is always accompanied by *three* points:

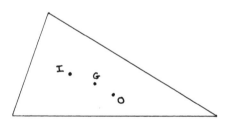

Why stop here? the altitudes (those perpendicular lines from vertices to the opposite sides) must also concur—it would be too strange if they did not. Given the way our story has evolved, you would expect that to prove this would be harder still. We're always wrong-footed by mathematics: it will take only the audacity of Alcibiades and looking askew to make this new truth appear.

The median proof in the Annex involved going down a tunnel inside a triangle; this one—to prove that the three altitudes are concurrent—reverses the direction. We'll take our triangle ABC and build another one *around* it.

The parallel postulate (once again vitally needed) guarantees that through C there is one and only one line ℓ parallel to AB—so let's construct it:

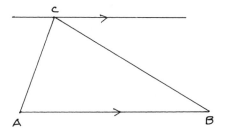

(our little arrows here mean that the two lines they are on are parallel).

Do the same now at A and B: through A, the only line parallel to BC, and through B, the only line parallel to AC:

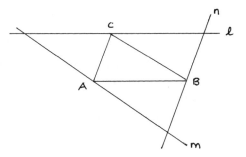

These new lines form a new triangle; we'll label its vertices R, S, and T.

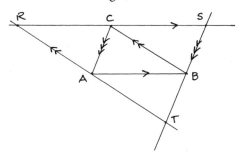

The ingenious person who first came up with this proof built such an enclosing triangle because it gave him two parallelograms, RCBA and CSBA (each is a parallelogram because in each, the sides are parallel in pairs). This guaranteed that RC = AB, and from the second parallelogram, that CS = AB. So by transitivity again, RC = CS, making C the midpoint of RS. You probably rightly sense that transitivity is as fundamental to our thought as triangles are to Euclidean geometry—that in fact it *is* the mind's triangle, showing us that going from one truth to another via a third means that we can now go directly.

If we chase the other parallelograms around in the same way, we see that A is the midpoint of RT and B of ST. Pretending that △ABC isn't even there and looking at ΔRST only, erect (at A, B, and C, of course) the perpendicular bisectors of the sides of ΔRST: w, x, and y. We know, from the very first theorem of this chapter, that they meet at a point: the circumcenter of ΔRST—but here, let's call this point H:

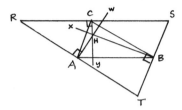

Of course, ΔABC won't go away, nor do we want it to. Since w is perpendicular to RT, it must also be perpendicular to BC, which is parallel to RT. x is, for the same reason, perpendicular to AC and y to AB. Yes—but this means that w, x, and y are the *altitudes* of ΔABC, and we have now proven them concurrent (at H), by thinking of them as lines serving another end in a different triangle. So the nimble mind coaxes new insights from old with that economy that marks the noblest arts. H is called the *orthocenter* of ΔABC, the fourth fixed point coded into every triangle's DNA.

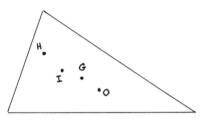

What we spoke of before as a minuet has turned out to be a quadrille: a quietly formal dance on the otherwise empty triangular floor. And what an intricate dance it is! Look again, for example, at a triangle ABC and its orthocenter H:

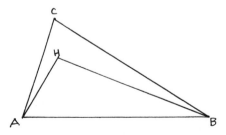

Draw AH and BH: then C is the orthocenter of ΔAHB! Why? Just turn your looking inside out: since an altitude is perpendicular to a side, the side must also be perpendicular to the altitude, and the two can switch roles in this masquerade. Is this an utterly unintuitive revelation, a tautology—or both? Take pencil to diagram to decide.

These stars shone singly in Euclid's sky. By the Age of Enlightenment they sang in glorious voice to Reason's ear, when Euler saw that the three points O, G, and H—the circumcenter, centroid, and orthocenter—are always collinear! The line they lie on is called the Euler Line. The proof has his easy genius to it.

ΔABC is either equilateral or not; if it is, O = G = H, so of course this one point is on a line. But if ΔABC isn't equilateral, then its centroid won't be its circumcenter, so draw the line from O to G and extend it twice its

Leonhard Euler (1707–1783), father of thirteen and endlessly productive in mathematics.

length to a point we hope will turn out to be H—so we'll call it H*.

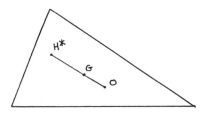

If we can prove that the altitudes all pass through H*, we will have proved that H* = H and so O, G, and H will be collinear.

First draw CG, and since G is the centroid, when we extend CG to meet AB at D, D will be the midpoint, since CGD is a median. And the perpendicular bisector will go up from D through O, since O is the circumcenter.

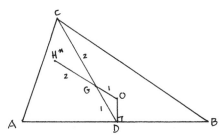

Because we know from page 115 that the centroid is $\frac{2}{3}$ of the distance from vertex to opposite side, we know that the ratio of CG to GD is 2 to 1. By the way we constructed it, that is also the ratio of H*G to GO.

The line begging to be drawn is from C to H*, continued to meet AB at K.

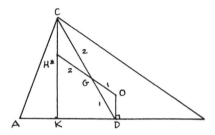

The little winged figure trapped inside ∆ABC is made up of ∆DOG and ∆CH*G, which have to be similar, since they have a pair of angles equal (the opposite angles ∠ CGH* and ∠ DGO), and the sides surrounding this angle are in proportion.

This means that since OD is perpendicular to AB, so is CK: so CK is the altitude from C, and it passes through H*. Reasoning similarly, the other altitudes show up passing through H*, so H* is indeed H, the orthocenter—which therefore lies on a line with the centroid and the circumcenter.

Once again Heraclitus is right: hidden relations are more powerful than those we see. These power-points of a triangle are subject to powers greater still.

Concurrent lines, collinear points—are there other fundamental shapes that hover invisibly over a triangle? Yes—and to call up a very surprising one we need only invoke one new figure to combine with those we already know: if you have a right triangle like FDN, then it fits neatly into—is *inscribed* in—a semicircle.

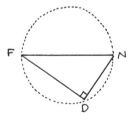

Long before Euclid was born, Thales proved that if a triangle is inscribed in a semicircle, then it is a right triangle—and sacrificed an ox to celebrate his discovery. So says Pamphile; and although she lived more than half a millennium later, it would be nice to believe her. It would be equally nice—and not that hard—to believe that Thales proved the converse too: "if a triangle is a right triangle then it can be inscribed in a semicircle"—for this follows in one step from our proof in the Appendix to page 108. Let's be generous and call this result "Thales's Converse".

Now we let our figures combine to recreate a discovery made by the reclusive high school teacher Karl Wilhelm Feuerbach in 1822: that in *any* triangle ABC, a seemingly random scatter of nine points—

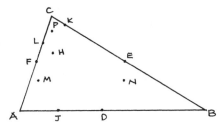

the midpoints of the three sides, the feet of the three altitudes, and the midpoints of the three line-segments connecting the orthocenter H to the vertices—must all lie on a circle!

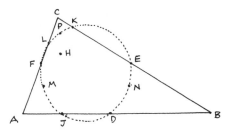

In our diagram these points are D, E, and F (the midpoints of the sides); J, K, and L (feet of the three altitudes, with orthocenter H); and M, N, and P, the midpoints of AH, BH, and CH, respectively.

The number of points involved and the late date of the discovery might lead you to suspect that the proof will be difficult; yet it uses no more than parallels and perpendiculars, parallelograms—and, as ever, transitivity.

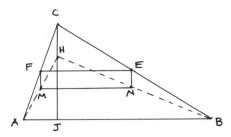

1. By our first result on page 114, FE ∥ AB (midpoints of sides in △ABC) and MN ∥ AB (midpoints of sides in △HAB); so by transitivity, FE ∥ MN.
2. Likewise, EN ∥ CH (midpoints of sides in △CBH) and FM ∥ CH (midpoints of sides in △CAH);
3. So by transitivity, FM ∥ EN.

4. This means that FENM is a parallelogram.
5. But since CHJ is an altitude, it is perpendicular to AB (CHJ ⊥ AB), so by transitivity again (and again), FM ⊥ MN and EN ⊥ MN.
6. That turns the parallelogram FENM into a rectangle.
7. Thales's Converse allows us to conclude that F, M, N, and E lie on a circle with diameter FN and center R, the midpoint of FN.

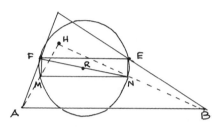

We are a third of the way there. The next part of the proof is exactly like the first, but looks at points F, D, N, and P. These too, and for the same reasons, are the vertices of a rectangle:

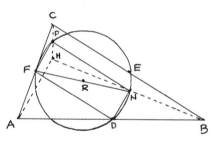

One of its diagonals is FN, so these four points F, D, N, and P lie on a circle with diameter FN and center R at its midpoint—the *same* circle, therefore, as before, so that F, M, N, E, D, and P all lie on it:

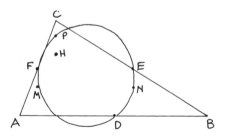

But what about J, K, and L, the feet of △ABC's altitudes? The diagonals of our two rectangles—FN, EM, and DP—are all diameters of this circle with radius R.

Look at diameter FN. The right triangle FLN is built on it (since BNL is an altitude, L is a right angle),

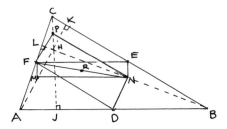

so by Thales's Converse once more, L lies on the circle with diameter FN and center R—our circle.

Likewise the right triangle PJD is built on diameter PD, so J is on this circle; and right triangle MKE is built on diameter EM, which means K is on it.

So all nine points lie on this single circle, called by some the Nine-Point Circle, others the Euler Circle—but most appropriately the Feuerbach Circle, especially since he also noticed that it is tangent to four other important circles: externally to the three circles tangent to the sides of the triangle, and internally to the incircle.

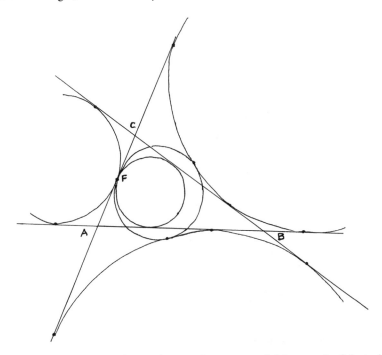

Will it surprise you to learn that R, the center of this wonderful circle, lies on the Euler Line? And would you be surprised to learn that this story is hardly over? For look at small triangles formed within the original one by taking each vertex with the two adjacent feet of the altitudes: each has, of course, its Euler Line (unless such a triangle is right or equilateral)—and these three lines are concurrent at a *tenth* point on the Nine-Point Circle. Great fleas have little fleas . . . (to see them, look at the Appendix).

Our brief glimpse at what seemed an empty triangle has uncovered five points, a peculiar line, and now this circle that always accompany it, invisible as familiars to all but those who know the spells to make them appear.

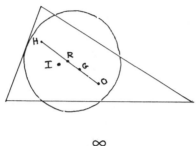

∞

        . . . stare
  At nothing, intricately drawn nowhere
  In shapes of shifting lineage . . .

Edna St. Vincent Millay wrote that in her sonnet "Euclid alone has looked on Beauty bare." Alone? Thales too, and Euler and Feuerbach, Hobbes and how many others, doodling on telephone pads, heard the call and learned how to look at this pregnant nowhere—or is it every-

Henri Poincaré (1854–1912)

where, these triangles that are only represented by diagrams but lie somehow behind or beyond or *within* them? Isn't geometry, as Poincaré once said, *"L'art de bien raisonner sur des figures mal faites"*— the art of reasoning well from ill-drawn figures?

These shapes seem so much more concrete than numbers do; yet just how elusive (remote and at the same time pervasive) they are, a last excursion will show.

What if you asked—as the amateur mathematician Count Giulio Carlo de' Toschi di Fagnano did in the 1700s— whether there was a triangle of shortest perimeter that could be inscribed in a given triangle: in effect, whether there is a least distance you could run and still touch each of the triangle's three walls.

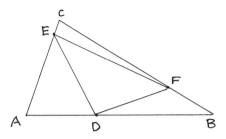

Would D to E, E to F, and F back to D be this shortest path?

The problem is interesting for many reasons. Logic says that of all possible paths there *ought* to be a shortest, but our intuition fails to tell us at once what the shortest path is—or even if there *is* a shortest (or if you are a formalist, whether the *existence* of such a path could be proved, even were the path itself not to be found). It is historically interesting also, because, like the two-faced Janus stones with which Romans once marked their frontiers, it looks both forward and back. Back, because "shortest distance" *always*, in Euclidean geometry, means "straight line", which, along with "point", is one of this geometry's two most primitive concepts. Ahead, because all questions about minimizing anything, such as a path length, are chiefly at home in the mathematics Euclid never dreamed of: calculus, which was the high point of seventeenth-century invention.

Yet how could there be any question of a straight line here, except for the obvious fact that each sprint across the triangle should be straight? The whole path couldn't possibly be. It is at this moment that the spirit of Alcibiades awakes at its most pugnacious.

"Shortest distance means straight line, and that is what I mean to have!"

"Ah, but you can't!"

"Yes, but I will!"

Half the mathematical insights that enlighten the world come from attending quietly to what the givens say; the other half—more riskily—from imposing your will on them. Young Alcibiades, playing in the dust, wouldn't move out of the carter's way and lived to tell the tale. Old Archimedes, continuing to draw in the sand when the centurion summoned him, didn't.

Let's go along with Alcibiades here and insist on a single straight line. It can't then be one that fits in the required triangle—but perhaps it could later be folded to fit, like a carpenter's rule. Yet how should we start? The answer also has an Alcibiadean cheekiness to it: start anywhere. Start with the D, E, and F of the last diagram. But how could that help? Those points were chosen at random; no chance of them having been the right ones—unless *every* path is equally short, which would be disappointing, or there is *no* shortest, which would be perplexing.

Keeping those two extreme possibilities in mind, our hope nevertheless is to hit on the one true way, and our strategy—following Polonius's advice to Laertes—is by indirections to find directions out. The idea is as clever as it is ancient: you will meet it even in the Rhind Papyrus, which the scribe A'h-Mose transcribed around 1650 B.C. from an Egyptian original some quarter millennium older. It acquired the name "false position" (*positio falsa*) in the Middle Ages; you could think of it as a tentative early adventure with x, the unknown.

The problem given in the Rhind Papyrus is to find a quantity such that when it is added to a quarter of itself, the result is 15. The method was this: choose any old number, and then adjust the result. Since we are free to choose, let's pick a number that will simplify our thinking: 4—because it is easy to find a quarter of it: 1. That would give us 5 (= 1 + 4) rather than the desired 15. Since 5 is a third of 15, our answer has fallen short by a factor of 3, so multiply the 4 we chose by 3, giving us 12, and behold! 12 when added to a quarter of itself, which is 3, yields 15. Any choice, of course, would have worked: had you chosen 2 instead, then 2 + $\frac{1}{2}$ falls 6 times short of 15, so you would have had to multiply 2 by 6— and so get 12 again.

The geometric equivalent of *positio falsa* is to choose (as we did) *any* points D, E, and F on the triangle's sides. Now for the first of two world-class insights: think of sides AC and BC as *mirrors* and reflect the point D in each of them, to X and Y, respectively, *outside the triangle*:

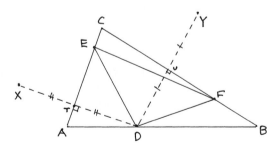

("reflect" means drawing a perpendicular from D to T on AC and then extending DT its own length to X—so X is the virtual image, through the glass, of D. Do the same thing with a perpendicular to U on BC).

Now connect X to our random point E, and F to Y, giving the zigzag path XEFY:

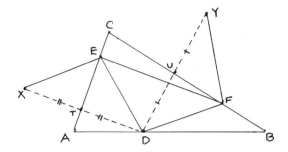

The reason for this bizarre maneuver is that XE is the same length as DE, part of our original, random, path: $\triangle TXE \cong \triangle TDE$ (by SAS: the shared side ET, the right angles, and the equal sides TX and DT), so EX = ED. Similarly, on the other flank, FY = FD. Hence XEFY is the same length as the path from D to E to F and back to D.

Suddenly we see how to satisfy Alcibiades's demand: if only XEFY were a *straight* line, it would be the shortest distance from X to Y—and therefore, so would the internal path it was reflected from. This means we can abandon two of our three random choices, E and F, and for the arbitrary point D get a shortest path as follows.

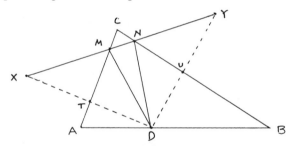

Reflect D in the "mirror" AC to X and in the "mirror" BC to Y; connect X and Y by a straight line. It will meet AC (at M) and BC (at N). Then D to M, M to N, and N back to D will be the shortest triangular path inscribed in the original triangle ABC—*if* we start at D.

Are we done? No, because although for a given D we now know where to find the other two points, we don't know where to station D along AB so that DMN will be the shortest of *all possible paths*. Or to put it in terms of a straight line: what choice of D will minimize the length XY?

This is where the second world-class insight appears (from what heaven of invention?). It wasn't Fagnano but Leopold Fejér whom the fiery muse visited. He taught in Hungary in the early twentieth century but almost didn't, his appointment having been opposed by anti-Semites on the faculty. One of them—knowing perfectly well that he had changed his name from Weiss—asked: "Is this Leopold Fejér related to our distinguished colleague in the Faculty of Theology, Father Ignatius Fejér?"

The eminent physicist Lóránd Baron von Eötvös answered at once: "Illegitimate son." Opposition ceased.

Copy Fagnano's construction with the falsely positioned D, and add the lines CX, CD, and CY:

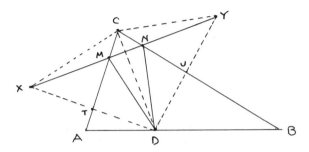

You might think of the see-saw XY hung by CX and CY from the balance point C: for just as before (looking at congruent triangles ΔTXC and ΔTDC on one side, congruent triangles ΔUYC and ΔUDC on the other), CX = CD = CY.

∠XCY can't change (it will always be twice the original ∠C, made up of ∠ ACD and ∠ BCD; and ∠ XCA = ∠ ACD, ∠ YCB = ∠ BCD), but CX and CY could shorten, in effect pulling up and shortening XY. How short can they get? Since each equals CD, it is just a question of when CD is shortest—and since the shortest distance from a point to a line is the perpendicular, this will be when CD is the *altitude* from C of ΔABC! So the inscribed triangle we first called DEF will have the least perimeter when our contrived line XY is shortest, and XY is shortest when D is the foot of the altitude from C.

Since there was nothing special about C and side AB, the same will be true of the other two sides: E must be the foot of the altitude from B and F the foot of the altitude from A: then D to E, E to F, and back again from F to D will be the minimal triangular path inscribed in ΔABC:

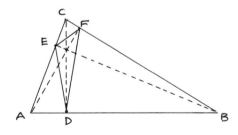

So the minimal path was built into the triangle's genetic code all along, a cousin of the altitudes whom we had just come to know. We discovered how to construct it by playing a game of *positio falsa* in our familiar old representative triangle, ABC.

Yet *how* representative was that triangle after all? A sudden doubt: will our construction work if the triangle is obtuse? For then some of the altitudes would meet not the opposite sides but their extensions:

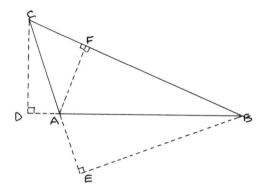

and clearly the path DEF fails to lie within △ABC. In fact, with F as one of the points on the path, where could the other two possibly be? For every choice such as D and E—

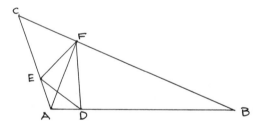

we could get a shorter path by moving D and E closer to A:

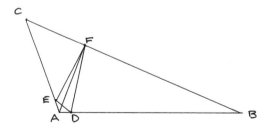

If we think of D and E as points on a number line, with A as zero, we know that for any choice we make we can always make a smaller—so there seem to be shorter and shorter but *no* shortest triangular path inscribed in an obtuse triangle (the path AF from the vertex A meets AC and AD at the same point, which only by an abuse of language fits our requirements). We have the same problem with a right triangle:

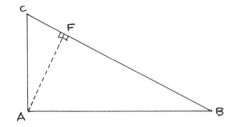

Since the feet of the perpendiculars are F and A, they give us no path save AF; and any other points chosen on AC and AB will slide together, approaching the single line AF as a limit, as they did in the obtuse triangle. This plausible argument isn't a proof but points to one, which you will find in the Appendix.

What we have just witnessed in solving Fagnano's Problem is an encounter found everywhere in mathematics: arrogance coming up against the natural resistance of things. The problem is solved for all time in any of the infinite kinds of acute triangle—solved by putting old objects in new arrangements. It is unsolvable for right triangles and the infinite varieties of obtuse triangles (or should we say: it is solved for these too, by knowing that there *is* no shortest path, unless you are willing to settle for a triangle with only one side?).*

"Arrogance" is what Alcibiades's enemies called what his friends saw as insouciant confidence. Intuition urges straight lines on a mind in pursuit of least distance, and reason has to contrive how to form intuition to fit the circumstances. Is this the way the two voices of Formalist and Intuitionist harmonize in us? What we first hear of their concord is this blithe, inventive tone. (A last marvelous example in the Appendix will tell us more about this tone and about what we once mistook for "innocent" triangles.)

How could the body's eye, which sees only what is, ever match the mind's, which also sees what might be? How could any particular diagram ever be adequate? The Formalist seems to have the last, cautionary word: if anything is infinite, it is the subtlety of the world. Yes, the Intuitionist answers, and of the mind (since each is a part of the other).

---

*Having seen that false assumptions may lead to false conclusions, we look much more suspiciously on the claims to generality of our old ΔABC. Should you take our word that the other properties we've proved are true for all triangles?

∞

Interlude

Longing and the Infinite

In 1187 the Frankish Kingdom of Jerusalem was attacked by the Saracens. Their leader, the Saladin, raised, they said, "an army without number, like the ocean." It was an army of 80,000 men. Nowadays we take that number in with ease: the crowd at a good-sized football stadium. There, of course, they have made their way through a turnstile and are sitting in ordered seats, not swarming toward us with spears. But we have our own versions of the countless: the midges at twilight, the sands on the shore, all the leaves on all the trees that were or are to be. Countless needn't mean infinite, just uncounted or hard to count.

Yet how easily our thought slides away from the very many to the infinite, as if we were anxiously eager to grasp infinity through an image. The eagerness is anxious, however, because an ancient interdict lies behind it: thou shalt not make graven images. Kenneth Clark explains:

> The voice that spoke to Moses out of the burning bush, or the single almighty being who spoke through the prophets, was infinite, and to give him finite shape in visible form was blasphemy.

Why do we insist that our god be infinite? Why would being bound in a nutshell world give us bad dreams? Why do we fear the cloister more than the agora? Longing and love have always "the expansion of infinite things," as Baudelaire wrote, and the distant beloved her infinite variety. The romantic in us wants always to be there, not here, at every possible here, just as heaven is for our reach exceeding our grasp.

Longing prolongs. The more remote the object of our desire, the more incomparable it seems. How could we even set about describing it? The numerical faculty in us proposes images remote and vast enough to be commensurable with our awe, not only because they have a sculptural purity to them but because the abstract calls up in us a tension between the distant and the near, very similar to what we feel for the remote be-

131

loved. You sense it in watching the prolongation of parallels to that ultima where they converge. It echoes in the series you see approaching its limit ever more closely, making "ever" itself thinkable and the infinite a diminishing fraction away.

Images grown abstract in geometry sidestep the Second Commandment, yet even in this rarified realm our infatuation with the infinite propels us beyond the geometric to forms that have no shape: the letters that stand for numbers, the numbers that stand for themselves. In its transparent doings, algebra echoes with angelic exchange—or are the promises algebra holds out to us diabolic (as the eminent mathematician Michael Atiyah suggests), since they may give us mere mechanism when what we wanted was meaning? To the algebraist, his timeless equations offer the prospect of understanding not just the infinity of past and future time, but the forms that hold time itself.

The totality of numbers strands midges and grains of sand at the starting line—yet through induction we plane over this totality and grasp its structure. Mathematics (the yearning halves of geometry and algebra completed to a whole) lets us see what Ravel detected in eighteenth-century French music: "Illimitable visions but of precise design, enclosed in a mystery of sombre abstractions." The abstractions of mathematics, however, are shot through with light, for it is to this art that all music aspires.

The Eagle of Algebra

Geometry sprawls as organically as a Gaudi apartment house, with Euclid's room tucked away down a corridor. Yet glancing through his door in Chapter Five we caught sight of astonishing vistas, each shaped from the simplest elements: points, lines, circles. When we suggested you draw one or another of these you probably didn't take us seriously, knowing we would do it for you—or more deeply, because mathematics lies in an enchanted world somewhere between reality and imagination. You *can't* make geometry's figures—only poor paraphrases of them: its lines have no thickness but infinite extent; its points have no dimension to them at all; its circles are perfectly rounded. Only the golden compass that William Blake's Ancient of Days holds in his hand at the world's beginning would suffice. Yet in our splodgy points and wobbly lines we know, without ever having seen them, just what is being represented, and recognize Triangle and Square themselves in our caricatures.

What does it mean, then, when Euclid asks us to *construct* a triangle, its circumcenter or circumcircle? What is being lifted up from those lines of Archimedes in the dust or of Hobbes on his thigh? Isn't it that we want to see how the gods (as the Greeks would have put it) would do this; or how (as we would say) our best efforts approximate ever more closely to a limit?

It is just here that another—and singularly Greek—consideration intrudes: the aesthetics of a frugal and seafaring people. If we can't get everything from nothing, let's try to get as much as we can from as little as possible. In this spirit the Pythagoreans, long before, compressed the harmonies of the universe into the tetractys. Our kit of tools for constructing should be as elegantly minimal as a mariner's. Euclid probably knew of subtle devices for making sophisticated shapes, but confined himself to an unmarked straightedge and compass. With these alone, he hoped to construct whatever came up in his geometry: certainly triangles and squares for a start, and why not five- and six-sided figures and in fact all the polygons that come after, whether they have 7 or 17 or 65,537 sides?

We looked for the basic generators of the Natural Numbers in Chapter Three and found the primes. In Chapter Five we saw that the triangle was the fundamental unit of plane geometry. Now the issue is actually to *build* anything in this geometry with Euclid's basic means.*

That issue may strike you by turns as insultingly simple and mind-bogglingly complex. Simple, because we know how to *picture* any such polygon—especially if we go one step further and ask only about those that are *regular*, in which all of the sides are equal. We can easily sketch a regular seven-sided polygon—a heptagon—for example:

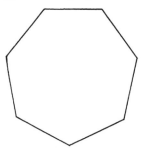

and filling in the cake slices from its center, can even say how many degrees must be in each such slice at the central vertex:

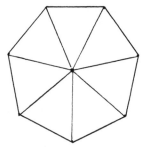

Since it takes 360° to circle around a point, the answer will be $\frac{360}{7}$ = $51.\overline{428571}°$. Given the dimensions of the slices we could calculate the heptagon's area and the length of its perimeter. Yes—but that's not the question. Can we in fact not sketch but *construct it exactly* with our two ideal tools?

This is where the mind begins to boggle: just how should we go about it, having no protractor? Even if we had one, the most delicate hand in the world couldn't capture the remoter digits of our endless decimal,

---

*Could one hope to go further? We have unfortunately not yet been able to read Juan Caramuel Lobkowitz's *Mathesis Audax* (1642), in which that Vicar General of England, Scotland, and Ireland seems to have resolved the major problems of logic, physics, and theology—above all, the issues of Grace and Free Will—by ruler and compass construction.

which the arithmetic mind so easily gauges. There must be a way, but it doesn't leap to the eye. Grown cautious over the course of the past five chapters, we may even want to reserve judgment about whether there always *is* a way. Mathematics seems ever to teach us two lessons: there is no limit to our mind's ingenuity; and there is even less of a limit to the intransigence of the world.

Let's begin our architectural work at the beginning and construct a regular (equilateral) triangle. Easy enough: since Euclid cares only for shape, not size, draw any length AB.

Now set your compass point at A, its pencil at B, and swing an arc:

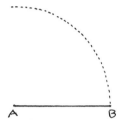

Reverse this process, putting the point at B, the pencil at A, and swing a second arc, meeting the first at C:

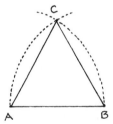

Now use your straightedge again to draw in line segments AC and BC. All sides, being radii of the same circle, are equal; hence ΔABC is equilateral. With such an easy beginning, the rest of the regular polygons should tumble to our will like induction's dominoes.

A square: from Chapter Five we know how to construct a line perpendicular to a line segment at its midpoint—call it B:

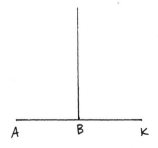

Now swing an arc with center B and radius BA, meeting this perpendicular at C:

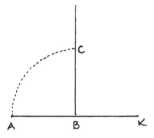

To conclude, swing two arcs with radius BA and centers A and C, meeting at D; then with your straightedge construct AD and DC, completing the desired square.

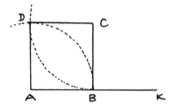

This playing off of compass against straightedge made triangle and square so easy to construct that you feel there must be something here that will generalize from n to n + 1. What, therefore, does it tell us about the pentagon? A deafening silence is all the answer we hear.

Let's make a strategic retreat and ask (as we did about the heptagon) how many degrees would have to be in each of its "central angles" α (Greek letters for angles—as a tip o' the hat to those who first told us about them, and to avoid confusion with Roman letters for points and lines):

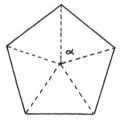

$\alpha = \frac{360°}{5} = 72°$. That looks more promising than what we got for a heptagon. If we could construct a 72° angle with straightedge and compass we could then iterate it four times around and so have our regular pentagon. (Since size doesn't matter, any circle from the center would put points equally far along each spoke, and we would make the sides with our straightedge between adjacent points.)

One of the high-risk appeals of mathematics is that you never know whether the next problem you stumble on might not lead you like Childe Roland to a dark tower. Constructing this angle of 72° will plunge us into a misty wood where everything seems a symbol instead of itself, and legends at least as old as the Pythagoreans call up intimations of mathematics as sorcery. It will take the eagle of algebra to rescue us from our amazement and bring us back to the greater light of how things intricately *are*.

It needs no more than drawing in the diagonals of our pentagon to discover within it the pentagram dear to the black arts. Look at the center of the pentagram: another pentagon! Draw its diagonals... Look at our original pentagon and a pentagram begins to take shape around it . . . tunnel one way, tower the other—moving in and out toward infinity.

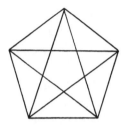

There is mystery enough just in our diagram: each diagonal divides whichever diagonal it crosses into a Golden Ratio. Euclid defines it this way: "As the whole line is to the greater segment, so the greater is to the less." We would now say:

$$\frac{1}{x} = \frac{x}{(1-x)} \ .$$

(We say it so casually, yet what a leap in thought this invitation to algebra involves, going one step beyond *positio falsa* and asking a letter to stand in for an unknown quantity, so that manipulating mere *forms* will reveal their content!) When the mask is lifted and we see the hidden number, this relation yields perfectly proportioned rectangles

which the far-sighted detect everywhere in the arts of their favorite culture (sides in a ratio of 2:3, 3:5, 5:8, 8:13, . . . better and better approximate this Golden Mean). Some musicians believe that the major sixth is the most beautiful interval because the frequencies it lies between are virtually in this ratio (or is it vice versa?). Since you can find it in shapes as different as nautilus shells and pine cones, enthusiasts of occult design discover it throughout animate nature and behind the Master Plan of Things. A significant (if not a golden) proportion of the Internet is devoted to its lore. Kepler, astride two worlds, wrote that "Geometry has two great treasures: one is the theorem of Pythagoras; the other, the division of a line into extreme and mean ratio. The first we may compare to a measure of gold, the second to a precious jewel." The Pythagoreans used the pentagram as a secret sign among themselves and called it "Health". Variants of the story we heard in the first chapter about Hippasus and his fate center around the pentagram rather than the irrationality of $\sqrt{2}$. The Tower of Mathematics, which is our frontispiece, derives, inverted, from Breughel's Tower of Babel. Its sides angled in, for him, at 72°—no doubt with referential intent.

How does any of this help in constructing the pentagon? We will leave it for a time and set off on a strategy at the heart of much of mathematics: let's call it "fetching from afar". Here the mathematician as Merchant Adventurer travels to realms remote from his problem in order to return enriched with the means for solving it. The golden ratio will be part of his cargo. You have seen this process at work before, as when we found a triangle's orthocenter by looking instead at the circumcenter of a different triangle.

Shuffling around in the attic of insight we come on this thought: were we able to construct a regular 10-gon (decagon), we would be able to construct our pentagon, simply by joining together the decagon's alternate vertices:

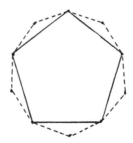

The central angle, α, of each slice of a decagon is $\frac{360°}{10} = 36°$, and since we are interested in regular decagons, the base angles of each slice will be equal, hence the familiar $\frac{(180°-36°)}{2} = 72°$ each.

138

If we take the side lengths to be one unit long, our task is to construct the segment forming the base. Call its length t.

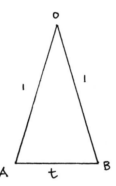

If we could do that, then we would draw a circle with center O and radius 1, choose any point A on its circumference and with radius t draw an arc intersecting the circle again at B. Doing this nine more times around the circumference would give us the points to join by straight lines, so making the decagon, and this in turn would yield our pentagon.

Yet how construct t? This is the moment no mechanism can rise to: only our prehensile minds. Imagine having our slice already constructed, and further imagine bisecting the angle at A, with a line meeting BO at D:

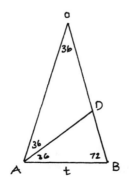

The 72° angle at A is now cut into two equal angles of 36°. ∠ADB is therefore $180° - (36° + 72°) = 180° - 108° = 72°$. If the base angles of a triangle are equal, so are the sides opposite them (by a proof identical in form to Pappus's daring contrivance in the appendix to page 130), so AB = AD: that is, both are of length t.

But ∠OAD = ∠AOD = 36°, so by the same theorem applied to ΔADO, AD = OD: OD is also of length t.

Since OB = 1, the segment DB = 1 – t.

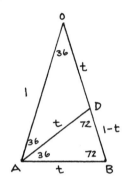

Here is our ship sailing home: ΔOAB ~ ΔDAB since their corresponding angles are the same. Hence their sides are in proportion:

$$\frac{\text{long}}{\text{short}} = \frac{\text{long}}{\text{short}}.$$

In this case,

$$\frac{1}{t} = \frac{t}{(1-t)},$$

the extreme and mean ratio!

Our task, however, is to construct the length t. How can we do this? It is now that algebra brings us its little formal touches, anticipated so long ago in Egypt, to free the unknown.

Multiply both sides of this equation by (1 – t) and then by t, turning it into

$$1 - t = t^2.$$

Collect all terms on one side:

$$0 = t^2 + t - 1.$$

Now if we could only solve this quadratic equation in t . . . (you'll find two ways of solving quadratics in the Appendix), we'd discover that

$$t = \frac{(\sqrt{5}-1)}{2}.$$

If we have a line segment, v, whose length is greater than 1 (and $\sqrt{5}$ *is* greater than 1), we know geometrically how to subtract a length 1 from it:

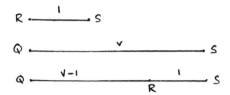

QR has length v – 1. So if we could construct a line segment of length $\sqrt{5}$, we could then construct another of length $\sqrt{5}$ – 1. And since we are masters of bisecting line segments, we could then make our segment t of length

$$\frac{(\sqrt{5}-1)}{2}.$$

∞

Hippasus Revisited

The Pythagorean world was shattered by Hippasus's proof that there were numbers, such as $\sqrt{2}$, which weren't the ratio of whole numbers. Horrified though Pythagoras must have been by the monstrous progeny of the simple straightedge and compass, might he (or we) not take comfort in the thought that regularity remained, but at one remove: the means for making these monsters—the Euclidean tools—were still ideally simple. A hidden regularity, revealed by reiterated applications of the Pythagorean Theorem, lies too among the offspring:

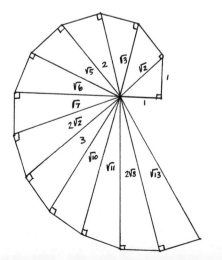

Harmony has been restored to the world—not on the level of its objects but of their making.

But what if for some other polygon you needed $\sqrt{\sqrt{\sqrt{17}}}$? A slick way to do the 2^nth root of any natural follows from similar triangles and Thales. For in a right triangle ABC we can drop the perpendicular to the hypotenuse (meeting it at D). Two new triangles are thus formed, similar to the original one and hence (by transitivity) to each other:

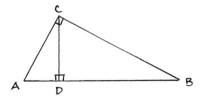

$\triangle ADC \sim \triangle ACB$ because each contains $\angle A$ and has a right angle. Likewise $\triangle ACB \sim \triangle CDB$ ($\angle B$ in common and the right angle). Their paired sides are therefore in proportion:

$$\frac{AD}{DC} = \frac{DC}{DB} .$$

Since we know from Thales's Converse (page 120) that a triangle inscribed in a semicircle is a right triangle, let's construct a circle of diameter $AB = 1 + 5 = 6$, and on this diameter place D so that $AD = 1$ (our given unit length). Then $DB = 5$.

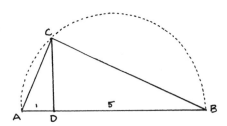

Erect a perpendicular to AB at D, meeting the circle at C; and draw AC and BC.

What is the length of DC (the *mean proportional* between 1 and 5)?

$$\frac{1}{DC} = \frac{DC}{5} ,$$

so $(DC)^2 = 5$ and $DC = \sqrt{5}$, as desired.

Now we can construct our way backward to the length

$$t = \frac{\sqrt{5}-1}{2} \; ,$$

hence to the regular decagon—

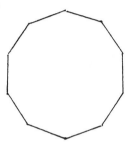

and therefore to the pentagon.

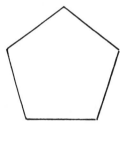

∞

The pentagon stretched our conceptual engineering. Will the hexagon be proportionally harder to construct? No: it takes hardly any work and no thought whatever, because we just fit six equilateral triangles around a central point: $6 \times 60° = 360°$, and equilateral triangles give us not only the central angles but the equal sides we need.

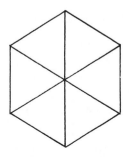

Suddenly it becomes clear that we can construct a 12-sided polygon (dodecagon) if we can bisect the central angles of the hexagon—

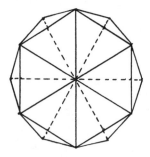

and indeed that we could have found the hexagon by bisecting the central angles of the triangle:

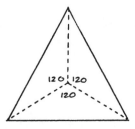

so if we could bisect an angle, every constructed n-gon would give us a 2n-gon for free—and angle-bisection falls readily to compass and straightedge. To bisect ∠AOB, swing any arc with center O, meeting AO at P and BO at Q.

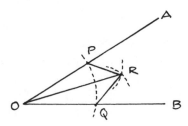

Then with radius PQ and center P, swing another arc, and do the same with center Q; these two new arcs meet at R. OR is the bisector of ∠AOB, since ΔORP ≅ ΔORQ: the paired sides are equal (a fourth way that Euclid establishes triangle congruence, called SSS).

What vistas this opens up! Now that we can bisect angles, the triangle will give us the hexagon, the hexagon the dodecagon, from that in turn the 24-gon, and so on—in fact, any member of the sequence $2^n \cdot 3$. The square gives us all polygons with $2^n \cdot 4$ sides, and now the pentagon all those with $2^n \cdot 5$ sides. Infinitely many regular n-gons are constructible, then—but the sophistication gained from Chapters Three and Four somewhat moderates our enthusiasm. 7, 9, and 15, for example, don't appear in any of these sequences, nor in fact do infinitely many others. The story may not be quite over.

Since $15 = 3 \times 5$ and we can construct triangle and pentagon, perhaps a little tinkering with them will give us the 15-gon. In a circle with center O construct a regular pentagon ABCDE (easily said, and now, with craftsmanship, done):

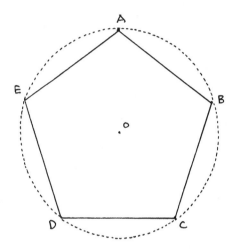

Starting at A, and with the circle's radius OA, move around the circumference marking off the points which would give the regular hexagon (see page 143):

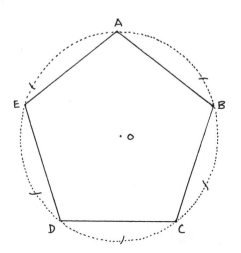

Now from A connect every other vertex, a new and easy way of making the equilateral triangle AFG:

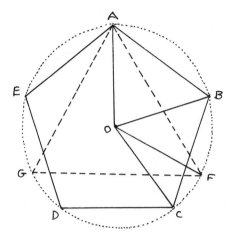

Draw OA, OB, OF, and OC. The central angle of the triangle, $\angle AOF$, contains 120° and the central angle of the pentagon, $\angle AOB$, contains 72°. So $\angle BOF = 120° - 72° = 48°$. $\angle BOC$ is also 72°; that means $\angle FOC$ is $72° - 48° = 24°$, the central angle of a 15-gon (since $24° = \frac{360°}{15}$). CF is therefore the side of a regular 15-gon. Setting the compass to length CF and swinging around the circumference will give us the rest of the 15-gon's vertices.

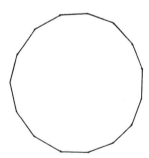

This technique will allow us to bring forth a new product-polygon from any pair of constructed polygons whose number of sides have no factors in common. We couldn't get 9 out of 3 and 3, for example, because the two triangles would merge into one. A square and a pentagon would give us a 20-gon, if we hadn't already lazily produced one through bisecting the angles of a decagon.

Even were we to go on constructing other hybrids, that wouldn't do away with the sort of irritation we've already felt when we were getting dribs and drabs of results about polygonal numbers, nor would it satisfy our impatience for the infinite. We need to be more serious; we want to be more systematic, and ask: precisely what can and can't be constructed with straightedge and compass? We want to join an Intuitionist's relish for making with a Formalist's delight in legal elegance. If a polygon requires lengths expressed by various kinds of numbers, can Euclid's tools really produce them? The way to find out is simply to build up the kinds of lengths they can construct. Our program is therefore to find out what kinds of things are constructible—to find a filter into which you can pour all the real numbers, which lets through only those (and all those) numbers that measure constructible lengths.

Let's start with something we know goes through the filter: the number 1, measuring a unit length. And then we have to be able to do arithmetic with it—adding it to itself any number of times to produce lengths which can be added together, subtracted from one another, multipled, and divided. We'll also need (as we saw in the pentagon) to construct lengths with square roots in them.

That unit length: since Euclid's structures, as you know, are utterly insensitive to scale, pick any length you like and grandly call it "1". Or, to satisfy a formalist, since Hilbert's axioms for Euclid guarantee the existence of three non-collinear points, choose any two of them to mark the unit on the line through them; then begin laying off, with our compass, this unit length head to tail again and again, marking points that stand for the natural numbers:

We can now add two naturals by laying out the second after the first, or vice versa:

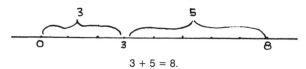

$$3 + 5 = 8.$$

Subtracting means laying off the length to be subtracted leftward from the head of the first:

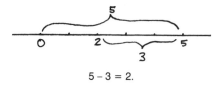

$$5 - 3 = 2.$$

Multiplication makes clever use of the fact that the sides of similar triangles are in proportion. We first saw it off-handedly at work in

Chapter One (on pages 14–15). To multiply 3 · 4, for example, with compass and straightedge, lay out four units on our number line; then on a line through 0 and that third point not collinear with our first two, draw a second line:

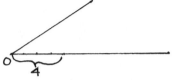

and on it lay off three units. For convenience, let's label the point marking zero A, the point at the end of those 3 units B, the point at 1 on the original line C, and the point at the end of 4 units, D.

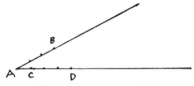

Since two points determine a line, construct the line through C and B, and at D draw a line parallel to CB, meeting AB at E*:

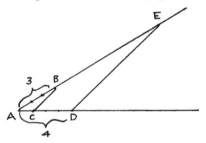

These parallel lines make ΔABC ~ ΔAED, so

$$\frac{3}{1} = \frac{AE}{4} \ .$$

Multiplying both sides by 4,

$$3 \cdot 4 = AE = 12 \ .$$

This same canny device leads to division and therefore to seeing ratios (for a Pythagorean) or constructing rational numbers (for us): a length of $\frac{3}{4}$, say, follows from this arrangement:

---

*The easiest way to construct a line ℓ′ through a point D, parallel to a given line ℓ, is to construct a perpendicular m from D to ℓ and then a perpendicular ℓ′ to m at D.

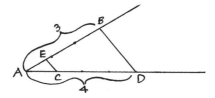

Here we draw BD rather than BC, then CE parallel to it. Since ΔACE ~ ΔADB,

$$\frac{AE}{1} = \frac{3}{4}.$$

That is, AE is $\frac{3}{4}$ of a unit long.

This means we can now locate any positive rational $\frac{a}{b}$ on our number line. In the bliss of this dawn, and for the sake of what is to come, let's continue our number-line (page 147) leftward from 0. The negative rationals will appear there as counterparts of the positives: $-\frac{a}{b}$ will be the same length away from 0 as $\frac{a}{b}$, but in the opposite direction. These new points, of course, mark negative numbers as *positions* on the endless number-line: we're not talking about negative *lengths*.

Peacock would have been pleased: whatever we could do with the natural numbers extends now effortlessly to the integers and rationals, all of which—through the Adam and Eve of straightedge and compass—obey every law for fields on Weber's tablets for Fields (page 38). We spoke of these numbers once as innocents in Eden but they seem more worldly here, standing about in their field like the folk of Piers Plowman's vision, which William Langland wrote five centuries before Weber.

> A fair field full of folk I found
> With all manner of men, the meaner and the richest,
> Walking and wandering as the world demanded.

This multitude we used to call call $\mathbb{Q}$ (the rationals); but with this vision in mind, let's rechristen it F, for field.

Our aim is to find out what can and can't be made with Euclidean tools. Where are the irrationals? We have seen so recently how to construct $\sqrt{5}$ with straightedge and compass, yet it is nowhere here, neither among those who, as Langland said, put them to the plough and practiced hardship in setting and sowing, nor with those who practiced pride and quaint behavior, and came disguised in clothes and features.

Let us invite them in, as the Old Masters would bring saints and angels (only a little estranged by their lighting and bearing) into mortal discourse on their canvases. We simply construct (as on page 142) a line-segment of irrational length—let's begin with that anchor of chaos, $\sqrt{2}$ —

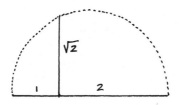

and drop it in amidst all the rational lengths of F. There let it go forth and multiply, divide, add, and subtract with all those established lengths and now with these new ones too, making every possible arithmetic combination. These will make up a new and larger field, which contains F as a subfield: a "square root extension field" of F, as it's called, and written $F[\sqrt{2}]$. Since this is our *first* field extension, we refer to it as F_1:

$$F_1 = F[\sqrt{2}].$$

Notice that closure, which we dismissed in Chapter Two as almost *infra dig*, turns out to be what matters here: F_1 is closed under all the arithmetic operations and square-rooting of 2 as well.

Every creature in this field will therefore have the two-part name $a + b\sqrt{2}$ —even though some may not at first seem to. "17" is $17 + 0\sqrt{2}$ in disguise; "$17\sqrt{2}$" is $0 + 17\sqrt{2}$ when it is at home. And $\frac{3+4\sqrt{2}}{1+\sqrt{2}}$? It takes a little clever encouragement to make it tell us its name. Multiply this quotient by $\frac{(1-\sqrt{2})}{(1-\sqrt{2})}$ and look what we get:

$$\frac{3+4\sqrt{2}}{1+\sqrt{2}} \cdot \frac{1-\sqrt{2}}{1-\sqrt{2}} = \frac{3+4\sqrt{2}-3\sqrt{2}-8}{1-2}$$

$$= \frac{-5+\sqrt{2}}{-1}$$

$$= \frac{-5}{-1} + \frac{\sqrt{2}}{-1}:$$

a is 5 and b is –1.

Although F_1 contains everything in F and an infinite number of other creatures besides—all of which we now see are constructible—$\sqrt{3}$ is not among them. Why not? Because $\sqrt{3}$, like $\sqrt{2}$, is irrational, so cannot lie in F. Nor can any arithmetic combination of rationals with $\sqrt{2}$ produce it (if in doubt, see the Appendix). That can't stop us, however, from building $\sqrt{3}$ in now. Since we know we can construct $\sqrt{3}$,

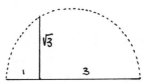

we act as we did before and adjoin it to F_1, to make the yet larger extension field $F_2 = F_1[\sqrt{3}]$. F_2, that is, has all the rationals in it, along with $\sqrt{2}$, $\sqrt{3}$, and all possible arithmetic combinations of these, with more or less obvious examples, like $\frac{3}{4} + \sqrt{3}, \frac{7\sqrt{2}}{19} - \frac{4\sqrt{3}}{13}, \sqrt{2} \cdot \sqrt{3} = \sqrt{6}, \frac{\sqrt{3}}{\sqrt{2}}$, which we know we can construct.

Since we can construct the square root of any already constructed number by the semicircle method, the program is clear. Whenever we find a number that was in a previous field but whose square root wasn't, adjoin this square root to the later field, just as we have done, to make a new field that will be the next link in our chain of fields—whose folk are constructible lengths. This means that if a number is in F or any square root extension field of F, then we can construct a line segment of that length with Euclidean tools.

How like the medieval notion of the Great Chain of Being this is! Any length in a square root extension field has been brought into existence by straightedge and compass. If it lies in F, the length is rational; if in F_1, it is an arithmetic combination of rationals and $\sqrt{2}$ (or as people say, $F_1 = F[\sqrt{2}]$); if in F_2, of these and $\sqrt{3}$ (that is, $F_2 = F_1[\sqrt{3}]$). We continue like this every time we find a rational whose square root is irrational yet lies in no previous field ($\sqrt{6}$ is irrational but belongs, as you saw, to F_2, since $\sqrt{6} = \sqrt{2} \cdot \sqrt{3}$; but $\sqrt{5}$, for example, requires a new link). In harmony with the medieval conception, this chain is infinitely long, since each prime has an irrational square root which—like $\sqrt{3}$—can't be derived from combinations of rationals with the square roots of other primes. In the language of a medieval metaphor, F begets F_1, which in turn begets F_2 which is $F_1[\sqrt{3}]$—and so on:

$$F \to F_1 = F[\sqrt{2}] \to F_2 = F_1[\sqrt{3}] \to F_3 = F_2[\sqrt{5}] \to \ldots$$

Is it awful or awesome that there are other links than these? For go back to F_1, containing all the rationals and all the arithmetic combinations with $\sqrt{2}$. Another length we could make, which isn't among them, is $\sqrt{\sqrt{2}}$, commonly called the fourth root of 2, or $\sqrt[4]{2}$.* We can construct it out of old material in the usual way:

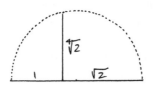

---

*Why is $\sqrt{\sqrt{2}} = \sqrt[4]{2}$? Whatever $\sqrt{\sqrt{2}}$ is, it is a number which, times itself, is $\sqrt{2}$. Four copies of it multiplied together will make $\sqrt{2} \cdot \sqrt{2} = 2$; hence $\sqrt{\sqrt{2}} = \sqrt[4]{2}$.

We therefore need a new link, which is F_1 with $\sqrt[4]{2}$ adjoined; and then that will call up another, since now we can construct $\sqrt[8]{2} = \sqrt{\sqrt[4]{2}}$, and then $\sqrt[16]{2}$, $\sqrt[32]{2}$, and in fact a link for each 2^nth root of 2, $\sqrt[2^n]{2}$. The same will be true for $\sqrt[4]{3}$, $\sqrt[8]{3}$, and so on, and the 2^nth root of any prime.

Our vision is turning nightmarish: infinitely long chains now hang down from each link of our infinitely long chain:

$$
\begin{array}{ccccccc}
F \rightarrow & F_1 = F[\sqrt{2}] & \rightarrow & F_2 = F_1[\sqrt{3}] & \rightarrow & F_3 = F_2[\sqrt{5}] & \rightarrow \dots \\
& \downarrow & & \downarrow & & \downarrow & \\
& F_{1,1} = F_1[\sqrt[4]{2}] & & F_{2,1} = F_2[\sqrt[4]{3}] & & F_{3,1} = F_3[\sqrt[4]{5}] & \\
& \downarrow & & \downarrow & & \downarrow & \\
& F_{1,2} = F_{1,1}[\sqrt[8]{2}] & & F_{2,2} = F_{2,1}[\sqrt[8]{3}] & & F_{3,2} = F_{3,1}[\sqrt[8]{5}] & \\
& \downarrow & & \downarrow & & \downarrow & \\
& F_{1,3} = F_{1,2}[\sqrt[16]{2}] & & F_{2,3} = F_{2,2}[\sqrt[16]{3}] & & F_{3,3} = F_{3,2}[\sqrt[16]{5}] & \\
& \downarrow & & \downarrow & & \downarrow & \\
\end{array}
$$

The bookkeeper closeted in every brain clutches his forehead and cries out, "How shall I ever arrange all these in order?" We'll mail him the astonishing directions in Chapter Nine. What matters here is that we don't require his skills: these fields needn't be stood to attention before our undertakings, but can be marshalled on demand to suit our needs.

Say, for example, that you have to construct an awkward length such as $745 \cdot \sqrt[32]{5} - (\frac{14}{3}) \cdot \sqrt{19}$. To start with, we know that 745 and $\frac{14}{3}$ lie in $\mathbb{Q}$, our base field F. Suiting our actions to our needs, let's first adjoin $\sqrt{5}$ to F, so that this time around F_1 will be F[$\sqrt{5}$]. We need to work our way down to $\sqrt[32]{5}$: $F_{1,1}$ will be $F_1[\sqrt[4]{5}]$, $F_{1,2}$ will be $F_{1,1}[\sqrt[8]{5}]$, $F_{1,3}$ will be $F_{1,2}[\sqrt[16]{5}]$, and finally $F_{1,4}$ will be $F_{1,3}[\sqrt[32]{5}]$. Now all we need do is adjoin $\sqrt{19}$, so this time $F_2 = F_{1,4}[\sqrt{19}]$—and it is in this F_2 that the required length can be constructed. What happened to $\sqrt{2}$ and $\sqrt{3}$, you might ask, and $\sqrt{7}$, $\sqrt{11}$, $\sqrt{13}$, and $\sqrt{17}$? We never needed them, and therefore built *this* chain of extensions from F without them. So a carpenter, with his templates and tools laid out in order, need only choose this one or that for the job at hand; he doesn't have to run through them all.

Lest you think, by the way, that *every* possible real number lies in some square root extension of F, notice that some don't: $\sqrt[3]{10}$ isn't rational (so it isn't in F) nor is it the square root, 4th, 8th, 16th, or any 2^nth root of any members in the square root extension chain from F. In fact, most of the cube roots of numbers lie outside our fields. This is true too of most 5th roots, 6th, 7th, 9th, 10th, and other roots not of the form 2^n—not to mention numbers like π, which aren't any sort of root at all. Populous though the links in our chain are, a vast array of numbers swarms outside them.

What matters to us, however, is that we have found our filter. Any length that lies in a square root extension field of F can be constructed.

But this criterion will help only if its converse is true as well: that any length which can be constructed lies in such a square root extension field. This is the second phase of our strategy and asks us to understand just what it means to construct with straightedge and compass. Fortunately the answer is clear: what we construct into existence are only the points where two lines or two circles, or a line and a circle, intersect. The particular points we have made thus far (in the pentagon, for example) have been so constructed—and have lain in F or one of its square-root extension fields (t—the side-length of a pentagon—is, as you remember from page 141, $-\frac{1}{2} + \frac{\sqrt{5}}{2}$, an element of that simple extension, $F[\sqrt{5}]$).

How can we be sure, however, that *all* the points made in one of these three ways will belong to a square-root extension field of F? Let's allow ourselves a luxury Euclid never had: the coordinate plane that glimmered in ancient Egypt and Greece, and that Fermat and Descartes brought fully to light in the seventeenth century. We simply set up a second number line perpendicular to the first, on each of which we can mark any point whose distance from the zero where they cross is a number belonging to F.

It is as if we had first moved from a map that showed only how to go eastward from home to one that extended westward as well; and here added a north and south to give us the world on a plane. Any point on this plane whose horizontal and vertical addresses both belong to F can now be located (and we'll always give the horizontal coordinate first).

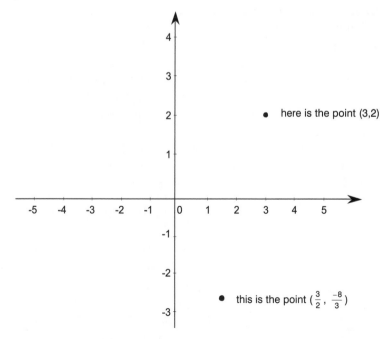

here is the point (3,2)

this is the point $(\frac{3}{2}, \frac{-8}{3})$

153

The words "plane" and "field" call up such similar images that you might think every point of the first belonged to the second. But keep in mind that the coordinate plane contains every single point that has real coordinates, while our square root extension fields are as exclusive as the parklands of great estates. You could think of it this way. The field of rationals, F, lies like a transparency on the Cartesian plane, with points all over it, corresponding to points with rational coordinates. $(\frac{3}{2}, \frac{-8}{3})$ is on it, but neither $(\frac{3}{2}, \sqrt{2})$ nor $(\frac{3}{2}, \sqrt[4]{5})$. The extension field $F_1 = F[\sqrt{2}]$ is a second transparency, with all the points of the first and now many more—all those that have at least one coordinate with a $\sqrt{2}$ in it. $(\frac{3}{2}, \sqrt{2})$ is here, but $(\frac{3}{2}, \sqrt[4]{5})$ is still missing. In fact, $(\frac{3}{2}, \sqrt[4]{5})$ won't be in *any* of the subsequent transparencies corresponding to links in the chain shown on page 152.

Nevertheless, the luxury of the coordinate plane will soon prove a necessity to us, and the power of algebra will lift us up above field after field, to see their ordered array. For it will let us find the *form* common to all points on a given line, and in particular a line through two points in one of our fields, and then the form common to all points on another such line. This will let us see the form of the one point on both lines—their point of intersection—and discover that it must have the form of a point in the field. We will do the same for a circle, then for its intersection with another circle or line built in the same field; and those points they have in common (their intersections) will turn out to be either in that field or in a square root extension of it. This will bring our strategy's second phase to an end, showing that our criterion was all we had hoped for: precisely those points that lie in F or some square root extension field of it can be constructed with Euclidean tools.

"The form of all points on a line": what does this mean? Not their *visual* form, which dots, no matter how small, approximate, so badly, but the form which is exact because abstract: their *numerical coordinates* (so far has our thought evolved from Chapter One). We want to be able to derive these coordinates from those of the two points the line was originally drawn through.

Take for example the line through the points (2,6) and (4,12). What form have the coordinates (x,y) of any point on this line in terms of 2 and 6, 4 and 12?

We notice for a start that this line rises steadily, with a constant slope—call it m—which is described by how far it moves vertically over a given horizontal stretch:

$$m = \frac{\text{rise}}{\text{run}} \, .$$

Since the vertical distance is the difference in the y-coordinates, and the horizontal the difference in x-coordinates,

$$m = \frac{(12-6)}{(4-2)} = \frac{6}{2} = 3 \, .$$

Hence the y-coordinate of any point on this line will be three times its corresponding x-coordinate:

$$y = 3x \, .$$

This particular line goes through the point (0,0). Any other line parallel to it must have the same slope, m = 3, but the y-coordinate of a point on it will be increased or decreased from y = 3x according to where such a line passes through the y-axis. The parallel line passing through (0,2), for example (2 units above our original line), will have points whose y-coordinates are given by

$$y = 3x + 2 \, .$$

The parallel through (0,–3) will give us

$$y = 3x - 3 \, .$$

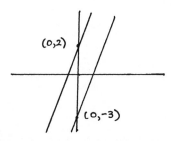

In general, then, the y-coordinate of any point (x,y) on a line with slope m, which intersects the y-axis at k, will be

$$y = \underset{\underset{\text{slope}}{\uparrow}}{m}x + \underset{\underset{\text{y-intercept}}{\uparrow}}{k} \, .$$

Now let's apply these results to *any* two points with coordinates in our field. Call these points (a,b) and (c,d). We can calculate m by taking the difference in y-coordinates (d − b) over the difference in x-coordinates (c − a):

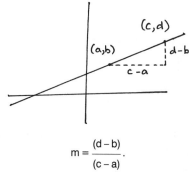

$$m = \frac{(d-b)}{(c-a)}.$$

You might be tempted to interrupt, saying that we're just lucky to have natural numbers for our initial coordinates: instead of a, b, c, and d we could have had rationals like $\frac{r}{s}$ in F, or hideous combinations in a square root extension field: a could have been $\frac{r}{s} + \frac{t\sqrt{2}}{u}$, and c, d, and e as bad. What an atrocious mess m would then be! But even were such intricacies lovely, dark and deep, the promise we have to keep is simply to show that two lines through points in a field intersect in another point of the field, and we still have some way to go. Benign neglect is called for here to avoid being sidetracked: a sort of blessed ignorance in which mathematics (which would know all things) thrives. What we care about is that m arises through some arithmetic combination of elements in the field; in this case, we have used subtraction and division. Let a, b, c, and d therefore stand for *whatever* those elements are; we need look no more closely in order to gain our end.

We now have

$$y = \left(\frac{d-b}{c-a}\right) \cdot x + k$$

↑   ↑

slope   y-intercept

and need to express k also in terms of our original four coordinates. We do this simply by turning the game around on itself (will the upcoming manipulations be exhausting or dreary? Neither: they afford the clock-maker's pleasure of watching the gears mesh). Since this equation puts y in terms of x for any point (x,y) on the line, it certainly does so as well for the original points (a,b) and (c,d). Choose one of them—say (a,b)— and in the equation above replace x by a and y by b, giving us

$$b = \left(\frac{d-b}{c-a}\right) \cdot a + k \ .$$

Solving for k with just a touch of algebra,

$$k = b - \left(\frac{d-b}{c-a}\right) \cdot a \ .$$

Hence

$$y = \left(\frac{d-b}{c-a}\right) \cdot x + b - \left(\frac{d-b}{c-a}\right) \cdot a \ .$$

We won't let these ugly expressions rattle us; the only message we want to carry away is that $b - \left(\frac{d-b}{c-a}\right) \cdot a$, like $\left(\frac{d-b}{c-a}\right)$ itself, is firmly within the field we started with. This means that given an x in that field, y (which is of course on the plane) will be in the *field* too. With this in mind, we can return with confidence to the more congenial form

$$y = mx + k \ ,$$

knowing that m and k are just arithmetic combinations of elements in our field.

We ask: if two lines arising from points of our field intersect, will their intersection lie in the field as well? Let that second line give the y coordinate of any point (x,y) on it by

$$y = \quad nx \quad + \quad g \ .$$
$$\qquad\qquad \uparrow \qquad\qquad \uparrow$$
$$\qquad\quad \text{slope} \qquad \text{y-intercept}$$

A firmer application of algebra will tell us the answer, since its aim is to extract the unknown from whatever circumstances it finds itself in. This is no trivial aim: "All the business of life," said Napoleon's conqueror, the Duke of Wellington, "is the endeavour to find out what you don't know by what you do; that's what I call 'guessing what was at the other side of the hill.'"

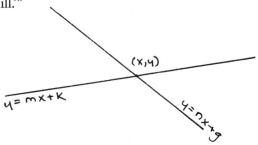

Our hill here has y = mx + k on one side of it, y = nx + g on the other. We are interested in the point (x,y) at the crest, where these two lines meet. It is the same point (x,y) on both lines, so that y = y; by transitivity,

$$mx + k = nx + g.$$

We want to find out the unknown, x, in terms of what we know: m, k, n, and g. Well,

$$mx - nx = g - k$$

so

$$(m - n)x = g - k$$

and

$$x = \frac{g - k}{m - n}.$$

This is just an arithmetic combination of elements in the field, so x must be in it as well; and we have already seen that if x is on a line derived from our field, so is the y coordinated with it. The point made by intersecting two lines of the field lies in the field too.

We've now found that the form of a line is approximately

———————————————————————

but is *exactly* y = mx + k.

We next need to find the exact form of a circle. Our thanks for this go back through Descartes and Fermat to Pythagoras, since his theorem holds the key to its equation.

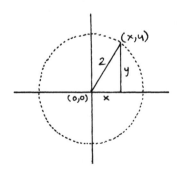

Say we have a circle of radius 2 with its center at (0,0) and want to know how the x- and y-coordinates of any point on it are related. As you (wearing the spectacles of Pythagoras) see in the drawing,

$$x^2 + y^2 = 2^2$$

that is

$$x^2 + y^2 = 4$$

or

$$y = \sqrt{4 - x^2} \ .$$

The adjustment is easy should the circle have radius r instead of 2:

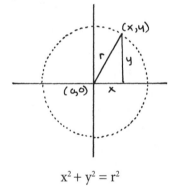

$$x^2 + y^2 = r^2$$

or

$$y = \sqrt{r^2 - x^2} \ .$$

The final modification displaces the circle's center from (0,0) to some other point (h,k) on the plane:

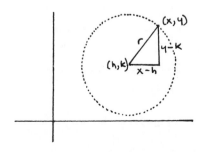

Now the circle's equation is

$$(x - h)^2 + (y - k)^2 = r^2$$

so

$$(y - k)^2 = r^2 - (x - h)^2$$

which gives us

$$y - k = \sqrt{r^2 - (x - h)^2}$$

or

$$y = k + \sqrt{r^2 - (x - h)^2} \, .$$

This is the algebraic form of a circle. It is the form latent in all the circles drawn in sand, on paper, or on your thigh. Without sweating any details we see that if x, r, h, and k lie in some square root extension field F_i, y will lie in at worst the next link from it.

Now we can take on the intersection of a circle and a line that both arise from some F_i. We hope that what points they have in common are in F_i too, or in a square root extension link from it.

Because the equation for a circle is more complicated than that of a line, the tactics for doing this will be more intricate than they were when we looked at the intersection of two lines—but the strategy remains exactly the same: to show that whatever happens, no more than arithmetic combinations and square rooting will be involved.

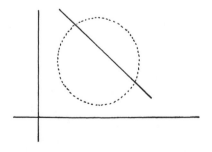

Our circle is $(x - h)^2 + (y - k)^2 = r^2$.
Our line is $y = nx + g$.

By transitivity (in the specific form of substituting the second expression for y into the first equation),

$$(x - h)^2 + (nx + g - k)^2 = r^2.$$

Whatever we will now do to free x from its entanglements, we won't go beyond adding, multiplying, subtracting, dividing, and taking square roots.

Like the sons of King Gama in Gilbert and Sullivan's *Princess Ida* who found their armor too heavy, we begin to remove the parts piece by piece. Squaring the two terms on the left makes things look temporarily worse:

$$x^2 - 2xh + h^2 + n^2x^2 + 2ngx + g^2 - 2knx - 2gk + k^2 = r^2;$$

but collecting like terms together,

$$(1 + n^2)x^2 + (-2h + 2ng - 2kn)x + (h^2 + g^2 - 2gk + k^2) = r^2,$$

we see that each of these expressions in parentheses is some arithmetic combination of elements in the field we began with, hence is in this field too—so call these three expressions A, B, and C, and off goes that helmet:

$$Ax^2 + Bx + C = r^2,$$

or simply

$$Ax^2 + Bx + C - r^2 = 0,$$

and since $C - r^2$ is also some element of the field—call it D—

$$Ax^2 + Bx + D = 0.$$

In steps the Quadratic Formula, and off goes that cuirass:

$$x = \frac{-B \pm \sqrt{B^2 - 4AD}}{2A}.$$

The same move repeated: $B^2 - 4AD$ is also some element in the field—call it E—so

$$x = \frac{-B \pm \sqrt{E}}{2A} \ .$$

x, stripped of its brassets and greaves, stands revealed in a square root extension field of the F_i we began with, namely $F_i[\sqrt{E}\ ]$.

If you have been dreading the final case—the points where two circles intersect—have no fear, but put yourself far enough above the battle to enjoy it; or succumb to the song of the sirens that invited Odysseus to their remote island:

> Here may we sit and view their toil
> That travail in the deep . . .

This travail is to solve "simultaneously", as timeless algebra so coyly puts it, the two equations

$$(x - h)^2 + (y - k)^2 = r^2$$

and

$$(x - j)^2 + (y - q)^2 = s^2$$

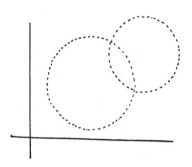

for the points (x,y) that are common to both.

Expand each equation, subtract the second from the first, carefully collect like terms together (as we did on page 161) and discover the form of a line hiding here:

$$y = \underbrace{\left(\frac{h - j}{q - k} \right)}_{\text{slope}} \cdot x + \underbrace{\frac{r^2 + j^2 + q^2 - s^2 - h^2 - k^2}{2(q - k)}}_{\text{y-intercept}} \ .$$

A *line*? Where did that come from?

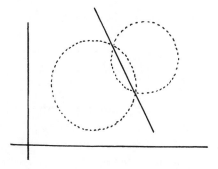

It is the "common chord" of the two circles, passing through their points of intersection; its constants (h, j, q, k, and their arithmetic combinations) lie wholly within the field we began with. So we are in the very situation we found ourselves in before, looking for the intersection of a line with a circle (here either circle)—and can confidently declare that we know those intersections will lie in that field or a square root link from it.

We are done. It has been like an exhilarating three-mile run—uphill. What we come away with is the certainty that the algebraic form of whatever we can construct with Euclid's tools has as its components only rationals and their 2^nth roots. You *won't* find cube roots, fifth roots, or such there (unless, exceptionally, one of those was a square root all along—as $\sqrt[3]{8} = \sqrt{4}$; or masquerades as a more complicated member of a square root extension field. So for example $\sqrt[3]{7+5\sqrt{2}}$ is $1 + \sqrt{2}$ in disguise). Now we see from aloft what we saw close at hand before: the pentagon *could* be constructed precisely because the lengths of the five sides involved nothing more exotic than $\sqrt{5}$.

∞

It was Gauss—once again Gauss, whose name runs through the last two centuries of mathematics like Louis Armstrong's through the evolution of jazz—who on March 30, 1796, when he was still eighteen, discovered how to construct the 17-gon. No one had seen a way, or was even sure that it could be done, in the two thousand years of thinking about it before him.

> By concentrated analysis I succeeded, during a vacation in Braunschweig, in the morning of the day, before I got up, to see [the general idea so] clearly that I was able to make the specific application to the 17-gon and to confirm it numerically right away.

A proof, if one is needed, that adolescents should be allowed to get up late during vacations.

We needed $\frac{\sqrt{5}-1}{2}$ to construct the pentagon. For the 17-gon Gauss needed

$$-\frac{1}{16}+\frac{\sqrt{17}}{16}+\frac{\sqrt{34-2\sqrt{17}}}{16}+\frac{\sqrt{17+3\sqrt{17}-\sqrt{34-2\sqrt{17}}-2\sqrt{34+2\sqrt{17}}}}{8}.$$

The $\sqrt{\sqrt{\sqrt{17}}}$ hidden here looks more frightening than Frankenstein's monster (which was born from Mary Shelley's pen only twenty years later). But if you look at the whole expression structurally, you see it belongs to some square root extension field not very far from F—and that is what matters.

It was this breakthrough that decided Gauss once and for all to become a mathematician (he had been equally attracted to philology, and wrote such beautiful Latin that some regret his nationalist friends, having persuaded him to write in German rather than the *lingua franca* of scholarship).

He published his result two months later:

> Besides the usual polygons there is a collection of others which are constructible geometrically, for example the 17-gon. This discovery is properly only a corollary of a not quite completed discovery of greater extent which will be laid before the public as soon as it is completed.

How near to glory we seem: for each of the regular polygons not yet constructed we need *only* show that its side-length lies in some square root extension field of F. Gauss tells us his construction of the 17-gon is *only* a corollary of a much greater discovery. You may think that this bell of "only" is pealing—but it is tolling: Gauss's discovery, which he published in 1801, was that in fact *not* all regular polygons can be constructed with Euclidean tools: he found that the only ones which *can*, have a number of sides of a rather peculiar sort. When you break this number down into its prime factors, each one will appear only once; moreover, these primes will have a striking family resemblance: each will be of the form $2^{2^k}+1$, for some natural number k. Of course once you construct a polygon you can—as we saw—construct another with twice as many sides, and twice again, and so on; hence some power of 2 will also be a factor (perhaps just $2^0 = 1$).* Any polygon whose number of sides doesn't fit

---

*Easy as doubling the number of sides is, it can have surprising consequences. You saw at the end of Chapter Four how an infinitely towering sequence of $\sqrt{2}$ s was miraculously equal to a finite 2. If you would like to witness a companion miracle, brought to you by the repeated doubling of a square's sides, turn to the on-line Annex.

this bill will have side-lengths whose equations will involve irreducible cube or higher roots, and so be unconstructible. What an astonishing and well-hidden unity behind the diversity of appearances.*

Which polygons *do* spring from primes of the form $2^{2^k} + 1$? The triangle: $3 = 2^{2^0} + 1$, and the pentagon: $5 = 2^{2^1} + 1$; and, as we know, any polygon with a repeated doubling of 3 or of 5 sides, or with $2^n \cdot 3 \cdot 5$ sides (such as the 15-gon, the 30-gon, etc.) can be constructed. Next comes Gauss's $17 = 2^{2^2} + 1$. And then? $2^{2^3} + 1 = 257$, which is prime. In 1832 two people named Richelot and Schwendenwein put in a little bid for immortality by showing how to construct the 257-gon.

$2^{2^4} + 1$ is also prime: it has in fact made several appearances in this book already, disguised as an arbitrary number no one would have thought about once, much less twice: 65,537. From Olympus a Mr. Johann Hermes delivered the construction of the 65,537-gon to the University of Göttingen, in 1879, wrapped up in a weighty manuscript written in the most admirable hand and filled with delicate drawings and cumbersome tables. It cost him ten years of his life and is there in its suitcase to this day, the most looked at and least read of all dissertations (as the curator of the University's collection of mathematical models remarks; you can see something of the puzzles it presents in the Appendix). How petty for any to scorn it as adding not a jot to progress. Do we dismiss the painstaking miniatures of the insane, or the ideal palaces built by provincial postmen? Where's your Forth Rail Bridge made out of toothpicks or your basement recreation of the Battle of Gettysburg now?

Fermat thought that all numbers of the form $2^{2^k} + 1$, called in his honor "Fermat Numbers", were prime. But $2^{2^5} + 1 = 4,294,967,297$ isn't prime (641 is one of its factors); nor is $2^{2^6} + 1$ nor $2^{2^7} + 1$ nor in fact any $2^{2^k} + 1$ for k from 5 to 32. As of May 28, 2001, 190 Fermat numbers, including this run from k = 5 to 32, have been checked, and none were prime. The largest number checked (you'll be flabbergasted to learn) is k = 382,447. To keep up to date, look from time to time at Wilfrid Keller's excellent web site: www.prothsearch.net/Fermat.html.

Is there another Fermat number out there which is prime? We just don't know, for we have suddenly arrived at a frontier of mathematics. If there are no more, then except for multiples of those we have, Hermes constructed the largest regular polygon possible. If there is a prime Fermat number $2^{2^k} + 1$ for some k beyond 32, no Hermes nor any Olympian will construct it in this universe, the number of whose particles to make *anything* with is significantly less than (the non-prime) $2^{2^{11}} + 1$. Yet in the infinite universe of the mind we may someday discover larger—or

---

*See the on-line Annex, too, for another surprise: a consequence Gauss drew from this criterion.

perhaps ever larger and larger—Fermat primes, or prove that, unlike the totality of primes themselves, these dwindle and die out, so that the species of constructible polygons are rarer than days in June.

What are we left with? The heptagon can't be constructed with straightedge and compass, although we can sketch it wonderfully well, see it approximated in hubcaps and coins, and seem to picture it perfectly with eyes closed. We may build a nine-sided city like Palmanova in Italy, but cannot construct its Platonic original.

We are left with a puzzle—it may even be a problem—about the -ible in "constructible": able how, when possible in theory but not in the physical world? Existing how, with singular points and special properties, when not even constructible theoretically? Embodied how, on the abstract Euclidean plane, when deposited there (as the heptagon is) by means less fundamental than Euclid's—such as marked straightedges slid along sophisticated curves? And does the ancient conviction echo here that thoughts are as real as or even more real than deeds (so that either might have been In the Beginning, and sinful thoughts now must as much be atoned for as sinful acts)? Or do constructions and constructing belong to the imagination, that messenger between the world and the mind, beholden to neither?

∞
seven

Into the Highlands

> The motto which I should adopt against a course calculated to stop
> the progress of discovery would be—remember $\sqrt{-1}$.
>
> —Augustus de Morgan

The Scottish chieftain Calgacus said of his country that it was "defended
by its remoteness and obscurity." The complex plane is the Scotland of
mathematics. The countryside we have passed through has been hilly,
but cities habitable for the mind of man have dotted it: the familiar integers here, triangles there (although once in them, the ways have often
turned mazy). Narrow all our roads down to the line of real numbers,
cross it with the line of the imaginaries, let it fade endlessly off in every
direction, and we are all at once in the Cairngorms.

What are we doing here? Many a climber has found that the little
chaos of life grows ordered and makes a new sense when seen from afar,
just as writers like James Joyce discover in exile the vivid structure of
home, concealed by its cluttered presence. Complex events in simple
contexts become simple when the context grows sophisticated. So on this
complex plane, exceptions and peculiarities, such as those we recently met,
will all at once be seen as outcroppings of deeper symmetries.

Simplicity and symmetry: how often the impulse toward understanding takes its bearings from these two markers, in the belief that ultimate
answers lie just beyond them (we lust after the subtle and singular as
openings into, rather than from, mystery). The complex plane promises
symmetry too, satisfying that old mathematical itch so well described by
William Rowan Hamilton:

> The algebraicist complains of imperfection, when his language pre-
> sents him with an anomaly; when he finds an exception disturbs
> the simplicity of his notation, or the symmetrical structure of his

syntax; when a formula must be written with precaution, and a symbolism is not universal.

Here is a striking sort of anomaly rectified on the complex plane. On the real plane, those quadratic functions we once had so much to do with come in three varieties. *Roots*—places at which the value of the function is zero—lie, naturally enough, on the x-axis, where y = 0. Some quadratics don't touch the x-axis at all, like $f(x) = x^2 + 3$; some at one place, like $f(x) = x^2 - 8x + 16$, whose only root is 4; and some in two places, and so have two roots, like $f(x) = x^2 - 5x + 6$, whose roots are 2 and 3.

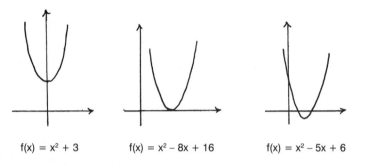

$f(x) = x^2 + 3$ \qquad $f(x) = x^2 - 8x + 16$ \qquad $f(x) = x^2 - 5x + 6$

Cubic functions can have one, two, or three roots, but the shape of their graphs forbids their having none.

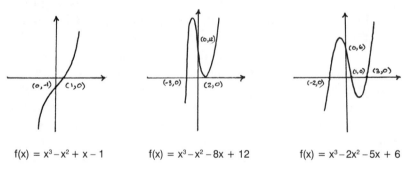

$f(x) = x^3 - x^2 + x - 1$ \qquad $f(x) = x^3 - x^2 - 8x + 12$ \qquad $f(x) = x^3 - 2x^2 - 5x + 6$

Quartic functions can have no, one, two, three, or four roots,

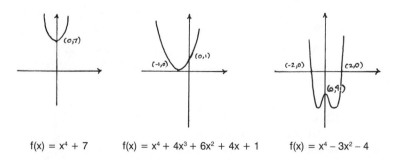

$f(x) = x^4 + 7$ \qquad $f(x) = x^4 + 4x^3 + 6x^2 + 4x + 1$ \qquad $f(x) = x^4 - 3x^2 - 4$

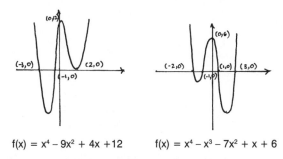

$$f(x) = x^4 - 9x^2 + 4x + 12 \qquad f(x) = x^4 - x^3 - 7x^2 + x + 6$$

and so on for higher degrees, needing a Linnaeus to classify them all. But if we allow *complex* roots, quadratics always have two, cubics three, quartics four—and nth degree polynomials always have n complex roots.* This truth (once again, proved by Gauss) is so important that it is called the *Fundamental Theorem of Algebra*. Roots are buried all over the complex plane, there for our extracting.

Is unreality the price we must pay for this tidying up? You have already heard the square roots of negative numbers called impossible as well as imaginary; but John Wallis, who had never studied math formally before he became Savillian Professor at Oxford in 1649, was less prejudiced. He saw a negative area as the spatial equivalent of a negative length: both represent loss. You can go into debt, and the sea can overwhelm your fields; and if a square has area −1600, it can only have a side of $\sqrt{-1600}$, or 40i. Let's see how they behave when the usual demands are made on them.

Adding is straightforward:

$$\begin{array}{r} 3 + 2i \\ + \underline{4 - 5i} \\ 7 - 3i \end{array}$$

so too is subtraction:

$$\begin{array}{r} 3 + 2i \\ - \underline{(4 - 5i)} \\ -\ 1 + 7i\,. \end{array}$$

The important point to notice here is that arithmetical combinations of complex numbers keep real parts with real, imaginary with imaginary.

---

*This is counting "multiplicities": if the same factor occurs twice, for example, in the polynomial, it is thought of as having two roots—or one root of multiplicity two—at that point. So $x^2 - 8x + 16 = 0$ has a root of multiplicity 2 at $x = 4$, since it is $(x - 4)(x - 4) = 0$ in disguise.

So in multiplying,

$$
\begin{array}{r}
3 + 2i \\
\times \ \ 4 - 5i \\
\hline
12 + 8i \\
- 15i - 10i^2 \\
\hline
12 - 7i - 10i^2
\end{array}
$$

but $i^2 = -1$, so $-10i^2$ is $-10 \cdot (-1) = 10$ in disguise, and our product is $12 - 7i + 10 = 22 - 7i$.

The complex numbers, then, remain closed under addition, subtraction, and multiplication. Might division suddenly force them to open into yet more fantastic forms? What if

$$\frac{3 + 2i}{4 - 5i}$$

were no longer of the form a + bi, where a and b are real numbers (one or both possibly 0)? Since it isn't at all obvious how to go about answering this, we should look back in admiration at Rafael Bombelli, strolling in the garden of his patron's Roman villa.

Our miniature view is of the mid-sixteenth century. A war has interrupted Bombelli's draining of the Pontine marshes. He is puzzling over having to find the three roots of certain cubics and announces that he has "found another sort of cubic radical which behaves in a very different way from the others." He has to make sense of expressions like $\sqrt[3]{2 + \sqrt{-1}}$, which are, he says, neither positive nor negative. Bombelli thought he had come on novel creatures; how was he to guess that they were the very imaginaries that Cardano, the mathematician whose work he so much admired, had wrestled with a generation before? What we call bi and –bi Bombelli calls "more than minus" (*piu di meno*) and "less than minus" (*meno di meno*). Names only, like "Unicorn" and "Gandalf", of creatures that don't exist? He too thought them merely sophistic until he began watching them combine ("More than minus times less than minus makes plus. . .")—as if antic figures, even more mysterious than J. B. Brown's, were materializing in the umbrella pines behind him and were then fixed there through the solidity of geometric proofs.

Seeing that bi and –bi always appeared yoked together in his calculations gave him his clue: since $(a + bi) \cdot (a - bi) = a^2 + b^2$—a *real* number—he took a quotient like our

$$\frac{3 + 2i}{4 - 5i}$$

and multiplied it by 1, in the guise of

$$\frac{4+5i}{4+5i}$$

(the same tactic we used on page 150 to reveal the true identity of $\frac{a+b\sqrt{2}}{c+d\sqrt{2}}$.
This would leave its *value* unchanged but convert its *form* to

$$\frac{(3+2i)}{(4-5i)}\frac{(4+5i)}{(4+5i)} = \frac{2+23i}{16+25} = \frac{2+23i}{41} = \frac{2}{41} + \frac{23i}{41} :$$

a perfectly good complex number.

Complex numbers, then, remain closed under all four arithmetic operations: $\mathbb{C}$ is a field. But square-rooting took us out of fields before. Perhaps here too the square root of a complex number will no longer be complex but something richer and stranger. Let's experiment with i itself and see if $\sqrt{i}$ lies beyond the complex numbers. In Alcibiadean spirit we'll bet that it *is* complex. When we roll the dice, either a contradiction will get the better of us, or we will win.

In its official form, i is 0 + 1i. Our claim is that

$$\sqrt{0+1i} = a + bi$$

for some real numbers a and b, which we want to find. We resort to the tactics familiar from Chapter One and square both sides:

$$0 + 1i = a^2 + 2abi + (bi)^2$$

or, since $(bi)^2 = -b^2$,

$$0 + 1i = a^2 - b^2 + 2abi .$$

Since 3 + 2i, for example, isn't 5 of anything, when two complex numbers are equal (as here), remember that the two real parts must be equal, and the two imaginaries must be also. We therefore have

$$0 = a^2 - b^2$$

and

$$1 = 2ab .$$

We want to solve these equations simultaneously, and to do this dip into the algebraist's bag of tricks.

Since $0 = a^2 - b^2$,

$$a^2 = b^2 \, ;$$

and that is only possible if b = a or b = –a.

In the second case, however, ab = a · (–a), which is negative, so 2ab would be negative, and couldn't equal 1; hence, we'd lose on that roll of the dice. We can only hope that we will succeed with the other possibility, b = a.

This means we substitute a for b in the second equation, and get

$$1 = 2a^2,$$

that is,

$$a^2 = \frac{1}{2}$$

or

$$a = \pm\sqrt{\frac{1}{2}} \, .$$

We can metamorphose this answer a bit:

$$\sqrt{\frac{1}{2}} = \frac{\sqrt{1}}{\sqrt{2}} = \frac{1}{\sqrt{2}}$$

and multiplying this last by $\frac{\sqrt{2}}{\sqrt{2}}$ for the sake of a rational denominator,

$$\frac{1}{\sqrt{2}} \cdot \frac{\sqrt{2}}{\sqrt{2}} = \frac{\sqrt{2}}{2} \, ,$$

so

$$a = \pm\frac{\sqrt{2}}{2} \, .$$

When $a = \frac{\sqrt{2}}{2}$, since b = a, b is $\frac{\sqrt{2}}{2}$ and we get the unlikely looking result

$$\sqrt{i} = \frac{\sqrt{2}}{2} + \frac{\sqrt{2}}{2}i \, .$$

Very dubious—but look what happens when we test it: if this creature really is the square root of i, then squaring it should give us i:

$$\left(\frac{\sqrt{2}}{2}+\frac{\sqrt{2}}{2}\,\text{i}\right)^2 = \left(\frac{\sqrt{2}}{2}+\frac{\sqrt{2}}{2}\,\text{i}\right)\cdot\left(\frac{\sqrt{2}}{2}+\frac{\sqrt{2}}{2}\,\text{i}\right)$$

$$= \frac{1}{2}+\frac{1}{2}\,\text{i}+\frac{1}{2}\text{i}-\frac{1}{2}$$

$$= \text{i}.$$

It does! And if a = $\frac{-\sqrt{2}}{2}$, then again, since b = a, b will be $\frac{-\sqrt{2}}{2}$ also. Is ($\frac{-\sqrt{2}}{2} - \frac{\sqrt{2}}{2}$ i) *another* square root of i?

$$\left(\frac{-\sqrt{2}}{2}-\frac{\sqrt{2}}{2}\,\text{i}\right)\left(\frac{-\sqrt{2}}{2}-\frac{\sqrt{2}}{2}\,\text{i}\right)=\frac{1}{2}+\frac{1}{2}\,\text{i}+\frac{1}{2}\,\text{i}-\frac{1}{2}=\text{i}.$$

Two square roots—just as the Fundamental Theorem of Algebra predicted.

Alcibiades's gamble has paid off: no contradiction, but instead two complicated as well as complex square roots of i stand revealed. The first person to see that *any* algebraic operation on the complex numbers left them closed was Jean le Rond d'Alembert in 1747—a man who, although his life was polarized, was convinced that all knowledge was unified. He had been abandoned by his unmarried socialite mother on the steps of St. Jean-le-Rond in Paris and raised by a poor glazier's family. His noble father later paid for his education, but d'Alembert kept his allegiance to his stepparents. If this timeless story leads you to think that now and then or here and there are the same, consider how strange past styles and customs seem to us: the work in which d'Alembert proved his result was his "Reflections on the General Cause of Winds".

A broader revelation comes with our two roots of i: their wholly unexpected (counter-intuitive?) form means that the terrain hasn't yet fully coalesced, having been—as a historian says of Virginia—an idea before it was a place. If you find yourself in a country "fained by Imagination" (Virginia as described by Sir Humphrey Gilbert), the solution is to let imagination do what fantasy cannot: focus in

D'Alembert (1717–1783)

on detail, so that we can end up navigating as confidently as we do in the reals.

Algebra helped geometry in the last chapter: here geometry will repay the debt. Since addition of complex numbers was straightforward, let's see what it looks like on the complex plane which we first saw on page 27, looking just like the real plane, but with the y-axis occupied by imaginaries. How did the ingenious Wallis come up with that image? He realized that i was the *mean proportional* between 1 and –1, because

$$\frac{1}{x} = \frac{x}{-1}$$

$$x^2 = -1$$

$$x = i$$

and therefore, like the mean proportional we constructed on page 142, should rise perpendicular to the real number line, halfway between 1 and –1.

We were adding (3 + 2i) + (4 – 5i)—but where *are* these two numbers? We have the point (3,2) standing for the *pair* "3 of the reals, 2 of the imaginaries"—

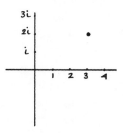

but how should we represent the *one* complex number 3 + 2i?

Once again, the simplest inventions often have the most profound consequences. In order to appreciate this one, savor the childhood revelation of one of our leading mathematicians, William P. Thurston. In the fifth grade he realized to his amazement that the answer to 134 divided by 29 was $\frac{134}{29}$. "What a tremendous labor-saving device!" he later wrote. "To me, '134 divided by 29' meant a certain tedious chore, while $\frac{134}{29}$ was an object with no implicit work. I went excitedly to my father to explain my major discovery. He told me that of course this is so, $\frac{a}{b}$ and a ÷ b are just synonyms. To him it was just a small variation in notation." Looking at one thing in two ways—here Euler simply set the two expressions equal: he let the point (3,2) on the complex plane *stand for* the complex number 3 + 2i. So small a step over so deep a chasm. Here then are 3 + 2i, 4 – 5i, and their sum, 7 – 3i:

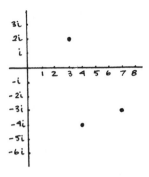

This picture doesn't seem to tell us anything. Try another: $(2 + 5i) + (8 + 3i) = 10 + 8i$:

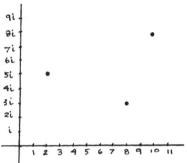

Again, neither Cassiopeia nor Orion shapes itself from these stars. Perhaps we have been spoiled by the constellations we found in Chapter Five, and the skeptic who walks in every optimist's shadow will rightly now step out into the sun.

It took a Norwegian surveyor to find the sight-lines. In 1797, Caspar Wessel—modest, self-taught, barely able to scrape a living from the maps he made of towns and coastlines and islands—published his paper "On the Analytic Representation of Direction; an Attempt". Why not think of these islanded points as the ends of arrows shot out from the origin, (0,0): directed line-segments, that is—or vectors, as we now call them. This is an idea that would come naturally to a sailor and chart maker thinking of the different forces of wind and current on a ship. An image begins to develop. Our first sum now looks like this:

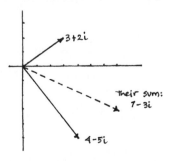

The same urge to symmetrize that we've felt again and again—the urge to complete the picture, the child's delight in connecting the dots—comes on us here: we sketch in the two missing lines that are longing to be found:

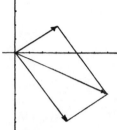

A parallelogram whose long diagonal is the sum! Has this homely shape, that played so important a part in Chapter Five, come to our aid far from home—or was it just a coincidence here? Examples may prove nothing but they do strengthen resolve, so let's try it again with (2 + 5i) + (8 + 3i):

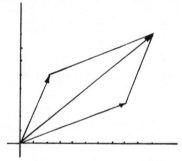

Once more it works! It *must*: adding (a + bi) to (c + di) means moving the first arrow, parallel to itself, a units over and b units up, so that its tail begins at the head of the second: and this gives us our parallelogram. Again, this is a notion congenial to anyone working with charts and the parallel rulers that transfer bearings from the compass rose to bearings from one's location.

And subtraction? Here, with Wessel's arrows, is (3 + 2i) − (4 + 5i) = −1 − 3i:

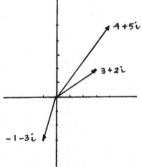

No parallelogram leaps to the eye. Yet something here is waiting to be born. If you draw the line connecting the first two arrowheads, it looks, oddly enough, parallel to and the same length as the arrow of their difference:

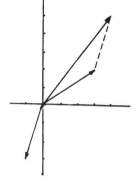

Perhaps this isn't so odd after all, if you think about what subtraction means: $(a + bi) - (c + di) = (a + bi) + (-c - di)$. Once we locate $-c - di$, our parallelogram incarnation of addition will give us the vector we want, with $-c - di$ the same length as $c + di$ but pointing 180° away from it. Hence the sum arrow of $(a + bi)$ and $(-c - di)$ will be parallel to the *other* diagonal of the parallelogram made from $(a + bi)$ and $(c + di)$:

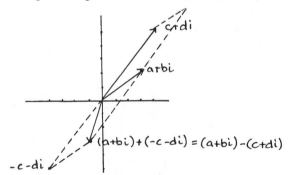

You might have thought that so stunning an insight as Wessel's would have been flashed around the world on the mathematical telegraph—had there been one. Instead, word from Norway languished in Scandinavia for a hundred years, during which time Wessel was knighted for his contribution to *surveying*. But in 1806, a self-taught Swiss bookkeeper named Jean Robert Argand rediscovered the idea (and so, inevitably, did Gauss in 1831). Why are these parallelograms now universally known as Argand diagrams? Perhaps because Argand's name came into such prominence when arguments raged over the validity of his figures. Servois—the man who coined the terms "commutative" and "distributive"—insisted that what was algebraic must be dealt with algebraically. The movement of Argand's thought from algebra to geometry, of Wessel's from geometry to algebra, shows once more how central to mathematical invention is fetching from afar (the analogue of metaphor in poetic invention).

We can now move about the complex plane as blithely as a summer visitor. How will multiplication look? $(3 + 2i) \cdot (4 - 5i) = 22 - 7i$:

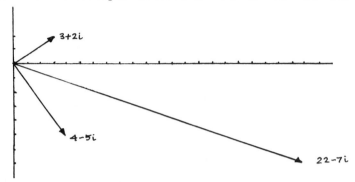

This is perplexing. Another example may shake our confidence further: $(2 + 5i) \cdot (1 + 2i) = -8 + 9i$.

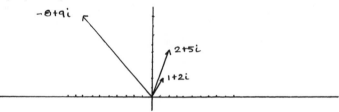

What is the product arrow doing so far away from those of its components? We seem to be faced with a truth we have confronted before: multiplication isn't some sort of shorthand for addition.

Now, however, we have accumulated enough experience to be sure that problems will have solutions—but to be sure as well that the way to them may be intricate. Finding the solution will show what multiplication "means"—and the intricacy of finding might make the pleasures of mathematics even more meaningful. For certainly what the twentieth-century mathematician Paul Halmos once said is true: "The major part of every meaningful life is the solution of problems." Not only is life, and the life of our imagination, thus enriched, but the world changes in ways we have yet to fathom. Hilbert once said: "There is the problem. Seek its solution. You can find it by pure reason, for in mathematics there is no *ignorabimus* [we shall not know]." Answering Hilbert's call brings into existence numbers no longer imaginary, and constructions that dovetail with those of ancient reality.

An important step in visualizing how complex numbers add was re-thinking the point (a, b) on the complex plane as a + bi, and then once again as a vector: an arrow from the origin. Yet another metaphor will carry the nature of multiplication across to us.

Look first, in our troubling diagrams for multiplication, at the *lengths* of the arrows. For 3 + 2i, the arrow is the hypotenuse of a right triangle:

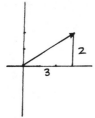

so its length is $\sqrt{2^2+3^2} = \sqrt{13}$. For 4 – 5i we have

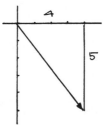

and the arrow's length is $\sqrt{4^2+5^2} = \sqrt{41}$.

The arrow of the product of 2 + 3i and 4 – 5i—namely, 22 –7i—

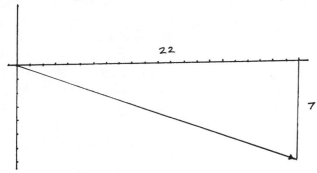

has length $\sqrt{22^2+7^2} = \sqrt{533}$. *

In other words, for the complex number a + bi the length of its vector is $\sqrt{a^2+b^2}$. This *real* number is called its *modulus*.

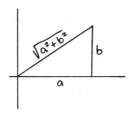

---

*Why does $\sqrt{13} \times \sqrt{41} = \sqrt{(13 \times 41)}$? Why, in general, is $\sqrt{a}\sqrt{b} = \sqrt{ab}$, if neither a nor b is negative? The full answer relies on Dedekind Cuts and how to multiply them. An example such as $\sqrt{4} \times \sqrt{9} = 2 \times 3 = 6$ and $\sqrt{4} \times \sqrt{9} = \sqrt{36} = 6$ makes it reasonable to expect that the general rule holds.

We have seen $a^2 + b^2$ before, on page 170. It is the number Bombelli came up with in making sense of division: the product of a complex number a + bi and its yoke-mate a − bi, called, therefore, its *conjugate*. What would Pythagoras have thought about his theorem reappearing to make sense of numbers so very remote from his own?

Now observe: $13 \times 41 = 533$: so $\sqrt{13} \times \sqrt{41} = \sqrt{533}$. For complex numbers, the modulus of the product is the product of the moduli! If this fails to reverberate harmoniously then look at this flow-chart:

$$(a+bi) \quad \cdot \quad (c+di) \quad = \quad (ac-bd) \quad + \quad (ad+bc)i$$
$$\downarrow \qquad\qquad \downarrow \qquad\qquad\qquad \downarrow$$
$$\text{modulus} \qquad \text{modulus} \qquad\qquad \text{modulus}$$
$$\sqrt{a^2+b^2} \quad \cdot \quad \sqrt{c^2+d^2} \qquad \sqrt{(ac-bd)^2+(ad+bc)^2}$$
$$\downarrow \qquad\qquad\qquad\qquad\qquad \downarrow$$
$$\sqrt{a^2c^2+a^2d^2+b^2c^2+b^2d^2} \quad = \quad \sqrt{a^2c^2+a^2d^2+b^2c^2+b^2d^2}$$

Half of our mystery is solved: we now understand—as Wessel and Argand and mathematicians like Euler before them did—the length of the product vector. But exactly where has this vector swung around to? Swung around: we can only come to grips with swinging in terms of angles. It was Euler who did this by wheedling from complex numbers the fourth of their names. He looked again at the line-segment from (0,0) to (a,b)—let's call its length r—and saw it as rotated counterclockwise from the horizontal by a certain angular amount ϕ (Greek letters once more for angles—this time phi):

The length r and that angle ϕ determine the segment's end-point as surely as do the coordinates (a,b), so he could now rethink a + bi in terms of r and ϕ:

$$a + bi = (a,b) = (r,\phi) .$$

We know how to derive the modulus r from a and b: $r = \sqrt{a^2+b^2}$. But how can we derive the angle ϕ? The way passes through the parkland of trigonometry (first cultivated by such Alexandrian mathematicians as Hipparchus, Menelaus, and Ptolemy two thousand years ago): a charming landscape, once you become familiar with its features. Here are a few pages from the guide to its flora and fauna. Keep in mind that our aim is to grasp the multiplication of complex numbers all at once: *seeing* it; and that angles will play an important role in this seeing.

The story is once again Pythagorean in spirit. As a line segment of a fixed length—let's simply make it 1—rotates counterclockwise from horizontal to vertical, it draws right triangles up with it, whose vertical sides grow in length from 0 to 1:

This is where sin enters math, as an abbreviation for sine (from the Latin *sinus*, for gentle curves from bend of bay to your brow's forecastle). The sine of angle φ, sin φ, is just the ratio of this opposite side's length to that of the hypotenuse:

$$\sin \phi = \frac{\text{opposite}}{\text{hypotenuse}} \ .$$

Since the hypotenuse here is 1, the opposite side's length in our triangle is just sin φ. So sin 0° = 0, sin 90° = 1, and sin 45° = $\frac{\sqrt{2}}{2}$, since both legs are equal and their squares add up to 1.

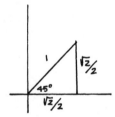

Any value of sin φ for φ between 0° and 90° can be figured out with more or less effort (your pocket calculator will do at once what cost men of the Renaissance, like Copernicus, hours and eyesight). The results produce a curving graph like this, when we relabel our axes from x and y to the angle φ plotted horizontally, and sin φ vertically:

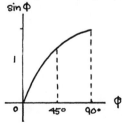

As φ goes on from 90° to 180° the side opposite φ decreases from 1 to 0 in the same way and at the same rate that we saw it grow:

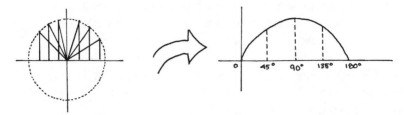

And if you attend to the plusses and minuses in the next two quadrants (180° to 270°, then 270° to 360°) and attach the relevant sign to the side-length, the graph of sin φ will go on to look like this:

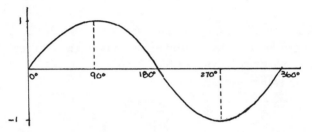

When you increase φ beyond 360° the pattern will repeat exactly (so sin 370° = sin (360° + 10°) = sin 10°, for example), giving us the sine waves that once dazzled adolescents on their basement oscilloscopes, before the Internet took them upstairs:

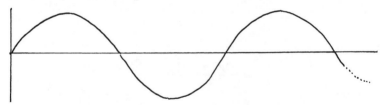

The side *adjacent* to φ will change as the opposite side did, but in reverse: shrinking from 1 to 0 as φ increases from 0° to 90°.

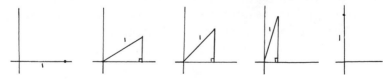

The ratio of this side's length to the hypotenuse is called cosine φ:

$$\cos \phi = \frac{\text{adjacent}}{\text{hypotenuse}} \ ;$$

so that here, where the hypotenuse is 1, the adjacent side is just cos ϕ. The graph of cos ϕ is the same shape as that of sin ϕ, but shifted left by 90°:

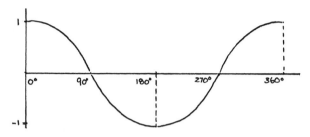

Looked at together, these two trigonometric functions braid perfectly:

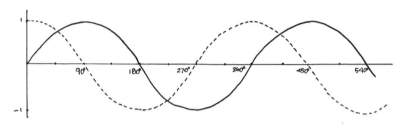

The braiding is even more apparent to the mind's eye focussed by Pythagoras:

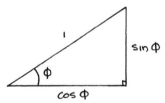

$$\sin^2 \phi + \cos^2 \phi = 1 .$$

Now we see how to relate the angle ϕ to our coordinates a and b on the complex plane: If the modulus is 1, a is just cos ϕ, and b is i sin ϕ:

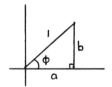

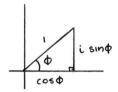

If the triangle is scaled up or down by a modulus r, each of its lengths is multiplied by r, and

$$a = r \cos \phi \qquad b = r \, i \sin \phi .$$

What was (a,b) is now (r cos φ, r i sin φ), so

$$a + bi = r\cos φ + r\,i\sin φ$$

or more economically,

$$a + bi = r\,(\cos φ + i\sin φ)\,.$$

"Mathematicians are like Frenchmen," Goethe once said; "whatever you say to them they translate into their own language and forthwith it is something entirely different." Nothing is sacred. Here they have even translated from one of their own languages into another.

We now have almost all we need in order to make visual sense of multiplying two complex numbers, a + bi and c + di, together. c + di will have its own modulus—let's say s—and its own angle, theta: θ. So

$$a + bi = r\,(\cos φ + i\sin φ)\,,$$
$$c + di = s\,(\cos θ + i\sin θ)\,,$$

and (a + bi) · (c + di) now becomes

$$r\,(\cos φ + i\sin φ) \cdot s\,(\cos θ + i\sin θ) = r \cdot s\,(\cos φ + i\sin φ)(\cos θ + i\sin θ).$$

Look! We see here what we saw before: the modulus of the product will be the product of the moduli. But what about those terms in parentheses? Carrying out the multiplication, being good about our bookkeeping and bearing in mind that $i^2 = -1$, we get the mantic

$$\cos φ \cos θ + i \cos φ \sin θ + i \sin φ \cos θ - \sin φ \sin θ\,.$$

Collecting real terms together at the front and the terms with i in them after, this becomes:

$$(\cos φ \cos θ - \sin φ \sin θ) + i\,(\cos φ \sin θ + \sin φ \cos θ)$$

so that altogether,

$$(a + bi) (c + di) =$$

$$rs \left[(\cos \phi \cos \theta - \sin \phi \sin \theta) + i (\cos \phi \sin \theta + \sin \phi \cos \theta) \right].$$

This is neater, but certainly not very neat; and no dazzling insight leaps from it to our minds. Beauty is truth, truth beauty, and both are mathematics. Something must be done about that clumsy, prowling quadruped.

The first thing to do is cage it. Let's take the triangle representing c + di, with angle θ, and move it temporarily to the real plane, so we can ignore the fact that its vertical side is in units of i, and call its length simply d. While we are at it, let's consider its modulus, s, to be 1. We'll bring back s and i after these simplifications have shown us the structure behind the symbols.

Now rotate the entire triangle counterclockwise by the angle φ belonging to the triangle for a + bi:

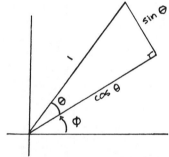

We'll want to refer to this triangle's sides from time to time, so label its vertices O, A, and B as here, and prop it up with a vertical line segment from A, meeting the x-axis at C.

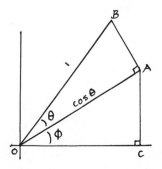

Finally, let's package our construction in a rectangular box:

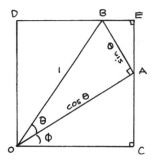

How long is AC? Since $\triangle OAC$ is a right triangle with hypotenuse $\cos\theta$, and $\sin\phi = \frac{\text{opposite}}{\text{hypotenuse}} = \frac{AC}{\cos\theta}$, solving for AC gives us

$$AC = \sin\phi\cos\theta,$$

and a tense stillness passes through our tiger.

By the same reasoning, $\cos\phi = \frac{\text{adjacent}}{\text{hypotenuse}} = \frac{OC}{\cos\theta}$, so

$$OC = \cos\phi\cos\theta.$$

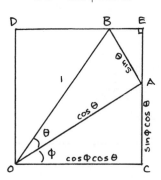

We need two more lengths: AE and BE. Since $\angle C$ is a right angle and $\angle AOC = \phi$, $\angle OAC = 180° - (90° + \phi) = 90° - \phi$.

But $\angle OAB$ is also $90°$, and since $\angle EAC$ is a straight angle ($180°$), $\angle BAE = 180° - ((90° - \phi) + 90°) = \phi$.

In $\triangle ABE$, therefore, $\sin\phi = \frac{\text{opposite}}{\text{hypotenuse}} = \frac{BE}{\sin\theta}$, so

$$BE = \sin\phi\sin\theta;$$

and $\cos\phi = \frac{\text{adjacent}}{\text{hypotenuse}} = \frac{AE}{\sin\theta}$, hence

$$AE = \cos \phi \sin \theta .$$

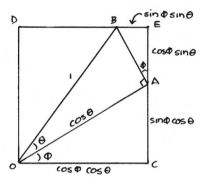

Why have we been playing musical chairs with these line segments? For the sake of our long-sought insight. If you now drop a perpendicular from B, meeting OC at F,

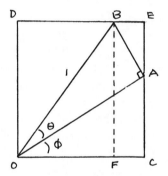

then $\triangle OBF$ has an angle at O of $\theta + \phi$, and $\sin(\theta + \phi) = \frac{BF}{1} = BF$, and $\cos(\theta + \phi) = \frac{OF}{1} = OF$.

But $BF = EC = \cos \phi \sin \theta + \sin \phi \cos \theta$, while $OF = OC - FC = OC - BE = \cos \phi \cos \theta - \sin \phi \sin \theta$: so that—gazing through the bars—

$$\cos(\theta + \phi) = \cos \phi \cos \theta - \sin \phi \sin \theta$$

$$\sin(\theta + \phi) = \cos \phi \sin \theta + \sin \phi \cos \theta .$$

When we substitute these telling expressions for their mute equivalents on page 185 we have:

$$(a + bi)(c + di) = rs [\cos(\theta + \phi) + i \sin(\theta + \phi)] .$$

The two terms added up in the brackets mean that to reach the point represented by $(a + bi) \cdot (c + di)$, we have swung through $(\theta + \phi)$ degrees and travelled rs from the origin. In other words, to multiply two complex numbers graphically, on the complex plane, multiply their moduli and add their angles!

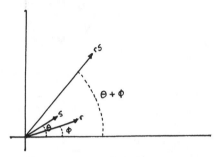

We have fought with demons of detail and triumphed. The sword of transitivity has worked its wonders. By passing from one way of expressing complex numbers to another, a key insight into how they behave has opened up. When Wittgenstein dismisses mathematics as nothing more than a string of tautologies, the mathematician answers: nothing *less*!

From the sixteenth through the eighteenth century, even the best mathematicians had used complex numbers covertly: as a means to be hidden or discarded before announcing the end they achieved. (In the same way, Archimedes apparently kept to himself his method of drawing mathematical insights from physical analogies, and Newton concealed under Euclidean geometry his radical ideas about calculus.) Even as late as 1825 Gauss wrote that "the true metaphysics of $\sqrt{-1}$ is elusive." Making the operations involving "imaginaries" visible, as we just did, gave them respectability at last. In 1831 Gauss wrote that their geometrical representation "completely established the intuitive meaning of complex numbers, and more is not needed to admit these quantities into the domain of arithmetic."

How can a meaning be established by a new representation, if it is already intuitive? We have suddenly spiralled back to the concerns of Chapter Two: what *is* this intuition which some appeal to as a court of first, others of last, resort? If common law can change, why cannot that of the intuition too? The numbers once stigmatized as impossible we now see behaving among themselves and the reals in a perfectly possible—in fact, cogent and attractive—manner, with a visual embodiment as well. Once again what was newfangled has become old hat, as habit fits its shape to our nature. Have we succeeded, then, in peeling off a layer that helped hide the endlessly deep core of our intuition—or only added one more colorful wrapping to an empty box?

∞

We have just made a long excursion into trigonometric functions in order to feel at home with the complex numbers. Before settling in to enjoy our hard-won discoveries, we would like to take one more excursion to a mountaintop where an astonishing view opens up. From it we will see that our new functions are the polynomials, familiar from Chapter Six, when spun out into infinite series like those we know from Chapter Four. Even more: the constant π, familar since childhood, will connect to e, that mysterious constant which lurks everywhere (surfacing momentarily in Chapters Three and Four)—and these two constants will be tied in a golden knot with i, as if in the welter we had caught a glimpse of unity.

"No great thing comes without a curse," said Sophocles. To reach this height we will have to avoid the gaping crevasse of calculus, as beautiful as it is deep, whose descent we *could* make had we the time. Instead we will follow the Greek precedent and set a sibyl over it, to speak oracles from its exhalations when we need them.

Like all good travelers we pack a bilingual dictionary in our knapsack. This one lets us convert the arbitrary degrees, with which we have up to now measured angles (only ancient arithmetic convenience, after all, broke circular measure into 360 equal shares), into the more natural *radians*, defined this way. Think of the radius as a short length of spaghetti, boil it for six minutes and you will find that you can lay it off along the curve of the circumference. Now since the circumference is $2\pi r$ long, precisely 2π radians (boiled radii) will lie around it. If we operate in our unit circle, where the radius is 1, our circumference will be 2π— that is, it will take 2π radians to complete the task we previously described as a 360° tour. π radians will take us halfway round, and $\frac{\pi}{2}$ radians will give us a 90° angle. In general, x degrees $= \frac{2\pi}{360} \cdot$ x radians. By convention, positive angles rise up from the x-axis; a picture will make all clear:

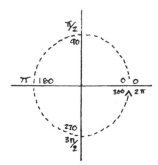

Not only are radians a more intrinsic measure of angles than degrees were, but they let trigonometric functions like $f(x) = \sin x$ or $g(x) = \cos x$ act as functions usually do, not with special "degree" inputs, but the normal real numbers that come from measuring distances around the circumference (reals between 0 and 2π if we go once around a circle, those between 2π and 4π if we wrap around it a second time, and so on). Negatives are defined as corresponding to angles measured *clockwise* from the x-axis.

A table of some outputs for sine and cosine will act as a rough guide to the region:

x in degrees	x in radians	sine x	cosine x
0	0	0	1
45	$\dfrac{\pi}{4}$	$\dfrac{\sqrt{2}}{2}$	$\dfrac{\sqrt{2}}{2}$
90	$\dfrac{\pi}{2}$	1	0
135	$\dfrac{3\pi}{4}$	$\dfrac{\sqrt{2}}{2}$	$\dfrac{-\sqrt{2}}{2}$
180	π	0	-1
225	$\dfrac{5\pi}{4}$	$\dfrac{-\sqrt{2}}{2}$	$\dfrac{\sqrt{2}}{2}$
270	$\dfrac{3\pi}{2}$	-1	0
315	$\dfrac{7\pi}{4}$	$\dfrac{-\sqrt{2}}{2}$	$\dfrac{-\sqrt{2}}{2}$
360	2π	0	1
405	$\dfrac{9\pi}{4}$	$\dfrac{\sqrt{2}}{2}$	$\dfrac{\sqrt{2}}{2}$

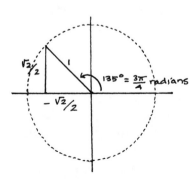

Indian mathematicians (probably before 1500), discovered wonderful infinite polynomial equivalents of sine and cosine and Newton redis-covered them independently in the seventeenth century. Here they are (the angle x is from now on measured in radians):

$$\sin x = \frac{x}{1!} - \frac{x^3}{3!} + \frac{x^5}{5!} - \frac{x^7}{7!} + \frac{x^9}{9!} - \cdots$$

$$\cos x = 1 - \frac{x^2}{2!} + \frac{x^4}{4!} - \frac{x^6}{6!} + \frac{x^8}{8!} - \cdots$$

The triple dots at the end of each line mean, as always, that the series continue in this pattern forever, with strict equality only after infinitely many terms. A few terms, however, give remarkably good approxima-tions. $\sin \frac{\pi}{4} = \frac{\sqrt{2}}{2} \approx 0.707106781$, for example, and the first five terms of the series for $\sin \frac{\pi}{4}$ yield

$$\frac{\pi}{4} - \frac{\left(\frac{\pi}{4}\right)^3}{3!} + \frac{\left(\frac{\pi}{4}\right)^5}{5!} - \frac{\left(\frac{\pi}{4}\right)^7}{7!} + \frac{\left(\frac{\pi}{4}\right)^9}{9!}.$$

Taking π as approximately 3.1415926535, $\frac{\pi}{4}$ would be .785398163, and five terms of our series would give us

$$\frac{0.785398163}{1} - \frac{0.484473073}{6} + \frac{0.298847348}{120} -$$

$$\frac{0.184344069}{5040} + \frac{0.113712689}{362880} = 0.707106782 :$$

only a few steps toward infinity give us an accuracy of eight decimal places!

What have these two series to do with **e**, that constant of exponential growth, which is approximately 2.718281828459045? We can raise **e** to various powers—even rational numbers and (with the help of calculus) any real number x, giving us a function

$$f(x) = e^x.$$

The infinite series equivalent of e^x was also discovered by Newton:

$$e^x = 1 + \frac{x}{1!} + \frac{x^2}{2!} + \frac{x^3}{3!} + \frac{x^4}{4!} + \cdots$$

Look at our three series together:

$$\sin x = \frac{x}{1!} - \frac{x^3}{3!} + \frac{x^5}{5!} - \frac{x^7}{7!}$$

$$\cos x = 1 - \frac{x^2}{2!} + \frac{x^4}{4!} - \frac{x^6}{6!} + \frac{x^8}{8!}$$

$$e^x = 1 + \frac{x}{1!} + \frac{x^2}{2!} + \frac{x^3}{3!} + \frac{x^4}{4!} + \frac{x^5}{5!} + \frac{x^6}{6!} + \frac{x^7}{7!} + \frac{x^8}{8!}$$

The mind reaches out a hand, longing to add the first two series in order to get the third—but the signs don't work out, with pairs of negatives after each pair of positive terms. In this cave of the sibyl, the ghost of Alcibiades calls out hollowly: "*Make* them work out!"

How? "Rely on your faith in pattern and readiness to see askew." But from what hills will the justification of this faith come?

"We are here in the Highlands of imaginary numbers: look to them."

This is what Euler did around 1740, experimenting with a mathematician's boldness. The functions sin x, cos x, and e^x are functions of a *real* variable x. But what if sense could somehow be made of putting in imaginary values, ix? Then since $i^2 = -1$, $i^3 = -i$, $i^4 = 1$ and so on, we would have

$$e^{ix} = 1 + \frac{ix}{1!} - \frac{x^2}{2!} - \frac{ix^3}{3!} + \frac{x^4}{4!} + \frac{ix^5}{5!} - \frac{x^6}{6!} - \frac{ix^7}{7!} + \frac{x^8}{8!} \ldots$$

He then regrouped these terms:*

$$e^{ix} = \left(1 - \frac{x^2}{2!} + \frac{x^4}{4!} - \frac{x^6}{6!} + \ldots\right) + i\left(\frac{x}{1!} - \frac{x^3}{3!} + \frac{x^5}{5!} - \frac{x^7}{7!} + \ldots\right).$$

In other words,

$$e^{ix} = \cos x + i\,(\sin x).$$

Amazing, and too good not to be true—and although it took more than a hundred years for others (such as Gauss and Cauchy) to make the sense Euler wanted of fitting in a complex variable where the real one had been, he was—like all mathematicians—easy with delay.

---

*You may recall from page 97 that we rearrange infinite series at our peril. If, however, a series converges when all its terms are positive, then we can legitimately rearrange its terms no matter how we change their signs. In the example on page 96, the series did *not* converge when all of its terms were positive.

Here was a reward for such insouciance. sin π = 0 and cos π = −1, as you can see in the picture:

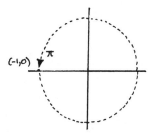

If we therefore let x = π in

$$e^{ix} = \cos x + i\,(\sin x)$$

we get

$$e^{i\pi} = -1\,.$$

Blink twice and look again: e, i, and π, those three remote peaks, have shimmered together to yield the barely more familiar mystery of −1. "Gentlemen," said Benjamin Peirce to his students at Harvard University one day late in the nineteenth century, "that is surely true, it is absolutely paradoxical; we cannot understand it, and we don't know what it means, but we have proved it, and therefore, we know it must be the truth."

∞

Now that we are at home in this land where hidden veins of gold surface, what pleasures and palaces will be ours! Look, for example, at the comfortable fact that any number will have two square roots. Those of 1, for example, are both itself and −1. But how many cube roots will it have? Since we are asking for the solutions of the equation

$$x^3 = 1$$

or

$$x^3 - 1 = 0\,,$$

the Fundamental Theorem of Algebra (stated on page 169) assures us that three answers are growing on the complex plane.

Three? Other than 1 again ($1^3 = 1$), what could they possibly be? We saw how unintuitive and downright ugly the two square roots of i turned out to be. Will we fare any better here? It seems not: solving by the method we used on page 171, the three cube roots of 1 we come up with are

$$1, \quad -\frac{1}{2} + \frac{\sqrt{3}}{2}i \text{ and } -\frac{1}{2} - \frac{\sqrt{3}}{2}i$$

(you may convince yourself of this awkward truth by carefully cubing each, or looking up the full story in the appendix for this chapter).

1 will have four complex fourth roots, five complex fifth, and so on—but if they are all going to be as unattractive as these, do we really want to meet them? The walking bass of our music has been: if it isn't beautiful, it isn't mathematics; and it sounds again through the overlaid voices here. The classical design of these "complex roots of unity" was uncovered by the work of a sequence of mathematicians, in which the first term was a man who correctly predicted the day of his death.

Abraham de Moivre was born in 1667 in France, fled to England when the Huguenots were expelled in 1685, and fell the further into poverty the higher he rose in the academic world. He studied annuities and mortality statistics; and perhaps with his thoughts so framed, noticed, they say, that he was sleeping each night fifteen minutes longer than the night before. From this arithmetical progression he calculated that on November 27, 1754, he would sleep for twenty-four hours, deduced that this would be the day he died, and did so. Prior to that he observed the immortal play of the complex numbers, all dressed in their new, trigonometric, finery. What he (and others after him, notably Newton's meticulous editor, Roger Cotes) saw was this.

The translation we made with much effort on page 187:

$$(a + bi)(c + di) = rs\left[\cos(\phi + \theta) + i\sin(\phi + \theta)\right]$$

takes on a nicely trimmed-down form when the two numbers are the same:

$$(a + bi)(a + bi) = rr\left[\cos(\phi + \phi) + i\sin(\phi + \phi)\right]$$

which is

$$(a + bi)^2 = r^2\left[\cos 2\phi + i\sin 2\phi\right].$$

Similarly

$$(a + bi)^3 = r^3\left[\cos 3\phi + i\sin 3\phi\right]$$

and in general,

$$(a + bi)^n = r^n [\cos n\phi + i \sin n\phi].$$

This has come to be called de Moivre's Formula. Stunning in its simplicity, it has a knockout consequence when we come to the roots of 1. If we want the two square roots (we know in advance that the answers are 1 and −1), we just recall that any complex number x satisfying $x^2 = 1$ will have the form r (cos ϕ + i sin ϕ) for an r and ϕ yet to be found.

Since 1 = 1 + 0i has the trigonometric form 1 · [cos 0 + i sin 0],

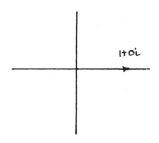

r = 1 and ϕ = 0

we must have

$$\{r [\cos \phi + i \sin \phi]\}^2 = 1 + 0i = 1 [\cos 0 + i \sin 0]$$

and de Moivre lets us rewrite the left-hand side, giving us

$$r^2 [\cos 2\phi + i \sin 2\phi] = 1 [\cos 0 + i \sin 0].$$

So $r^2 = 1$ (just as with real and imaginary parts, real moduli and these complex coordinates do not intermingle). If $r^2 = 1$, the modulus r = 1, because lengths can't be negative. And if

$$\cos 2\phi + i \sin 2\phi = \cos 0 + i \sin 0 ,$$

then 2ϕ = 0, or any equivalent of 0 radians as we wrap around the circle again and again: 0, 0 + 2π, 0 + 4π, 0 + 6π, . . . : in general, 0 + k · 2π radians, where k is a natural number. So

$$2\phi = 0 + k \cdot 2\pi ,$$

hence

$$\phi = \frac{0}{2} + k \cdot \frac{2\pi}{2} = k\pi,$$

for any natural number k.

When k = 1, $\phi = \pi$, and we get the −1 we expected:

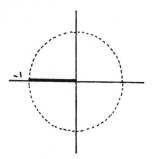

when k = 2, $\phi = 2\pi$, which thus takes us to the other square root of 1, namely 1:

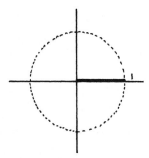

What about k = 3, 4, and so on? $3\pi, 4\pi, 5\pi, \ldots$ just keep taking us back and forth between these two square roots of 1: −1 and 1.

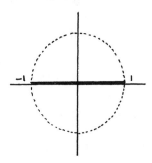

Interesting: they lie at opposite ends of this diameter.

What then of the three cube roots: will de Moivre help us transform the frogs of page 194 into handsome princes? Let's just repeat what we did with square roots. The complex numbers x satisfying $x^3 = 1$ have the form $r[\cos \phi + i \sin \phi]$, with now

$$r^3[\cos 3\phi + i \sin 3\phi] = 1[\cos 0 + i \sin 0]$$

once again the modulus $r = 1$, but now

$$3\phi = 0 + k2\pi$$

so

$$\phi = \frac{0}{3} + \frac{k2\pi}{3} = \frac{k2\pi}{3}.$$

For $k = 0$ we get 0 radians: the perennial root 1.
For $k = 1$, $\frac{2\pi}{3}$ gives us an angle in the second quadrant, and for $k = 2$, $\frac{4\pi}{3}$ an angle in the third quadrant.

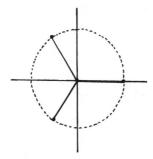

$k = 3$ yields 2π again (the root 1 we have already), and from 3 on we will only cycle through the roots already found.

Our geometric instinct springs awake: the three distinct cube roots of 1 are the vertices of an equilateral triangle!

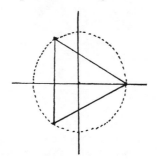

And the four fourth roots (oh, of course: 1, i, –1, –i) lie at the vertices of a square:

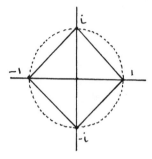

the fifth, sixth, seventh roots at the vertices of pentagon, hexagon, heptagon—

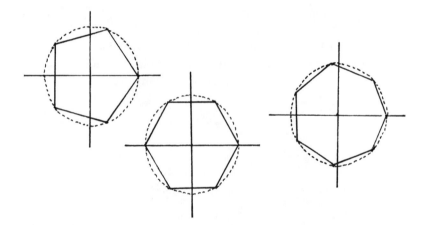

each of the regular n-gons, in fact, is reincarnated by de Moivre's Formula as an unexpected bearer on the complex plane of the n nth roots of unity.

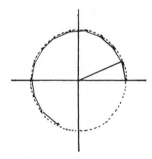

The n nth roots of unity, making angles of $\frac{2k\pi}{n}$ from the horizontal root at 1, for each k from 0 to n – 1.

On this plane we can even prove what we only asserted before—that the heptagon, for example, can't be constructed with straightedge and compass (see the Appendix).

All the peculiarities of the previous chapter dwindle away in this new dawn. There Euclid sought simplicity via his compass and straightedge, yet these led to intricate patterns with Fermat numbers, still tinged with mystery. Here the regular polygons make easy sense of puzzling complex roots. The way of mathematics is always to spiral up its widening tower to greater generality and higher simplicity. At these heights new objects and old interact and so give each other a different order of reality from that conferred by construction. For the ultimate atoms of mathematics are relations, not things, which therefore become more vivid the more they interplay. William Blake wasn't always right. "To Generalize," he said, "is to be an Idiot. To Particularize is the Alone Distinction of Merit." Yet he was right to continue: "All Sublimity is founded on Minute Discriminations." What singles out mathematics is the way that its minute discriminations lead to ever greater generalities, as climbers reach their dazzling vistas by attending to the piton in the cleft.

∞

Interlude

The Infinite and the Unknown

Mystery stories leave a flat aftertaste, because before the solution, anyone might have done it; after, it turns out to have been only a certain someone. But the infinite and the unknown endlessly call each other up, letting imagination loose.

We love to live on frontiers that enclose a polite, finite world but look out toward the ever unexpected. Is this mere romantic exuberance or the tinkering curiosity of our kind carried to its inevitable extreme? Whatever the cause, how refreshingly courteous of the world to oblige, always playing the tortoise to our Achilles by keeping a step ahead of all we ask. The hook of a question mark can't but snag more than it can bear.

Yet why should this be, especially if Spinoza was right in saying that the order of the world and the order of the mind are the same? It must involve a deep trait of our thinking, that no sooner do we make sense of this or that hang of things then all the intricate net shrinks down to a knot, just as a word comes to condense great stretches of feeling and event. That knot now sits among other abbreviations, demanding new ties among them in a more rarified net: and this is where our renewed explorations take place.

So when the universe seems to conclude in a Theory of Everything, a window will open up in the far wall onto a landscape unguessed at until then. Paradigms busily tidy up their last details just before they shake and shift.

Here we have seen the vivid complexity of triangles shrunk to no more than a point among the vaster collection of polygons which has its own ecology; and polygons are in two dimensions what polyhedra are in three and polytopes in four dimensions, and limitlessly beyond, O brave new worlds! And this unknown we step into is at least partly of our own making.

Mathematics builds upward by taking as stones what were structures before, gaining new heights from which to survey the way things are. How staggeringly far it has come in five thousand years—but for every answer found, a flurry of new questions arises. In the sequence of ratios of what we don't yet know to what at any moment we do, the terms grow to infinity.

eight

Back of Beyond

This has been the romance of imagination and the infinite. Like the beloved in tales as old as time, the infinite keeps escaping imagination's stratagems, drawing it on through intrigues that must any moment surely untangle. Mathematics being the stuff of invention and mathematicians each Alcibiades in disguise, why not just declare (since faint heart never won fair lady): here is the infinite, right here, in your midst. You have only to recognize it to make it yours.

Easily said, but how is it to be done? Think of Euclid's plane everywhere stretching away, with its parallel lines that meet at no "here" you can picture—unless it be through Alberti's Veil.

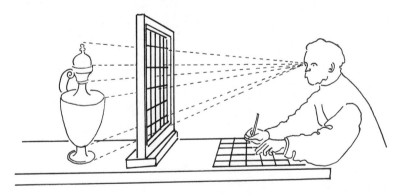

That wonderful Florentine, Leon Battista Alberti, shared the Renaissance eagerness to translate the beauties of the visible world into painting—to represent its depth on the plane—and saw how to do it by making "a veil, loosely woven of fine thread, dyed whatever color you please, divided up by thicker threads into as many parallel squares as you like, and stretched on a frame. I set this up between the eye and the object to be represented, so that the visual pyramid passes through the loose weave of the veil."

This notion of the visual pyramid (we might say "visual cone") was the key for turning three dimensions into two—a pair of pyramids, really, with our eye at the near apex, the "vanishing point" ordering pictorial space at the far, and the veil in between changing one image into another:

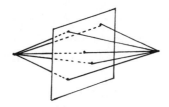

What was that vanishing point if not where the parallels receding from us palpably met?

The principles of perspective drawing developed with Italian gusto. How, for example, should a receding tiled floor be correctly drawn? Alberti's answer was ingenious: space the lines of the receding edges equally far apart:

then put the nearest pair of *horizontal* edges where you will,

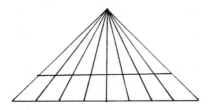

draw a diagonal through the left-handmost tile, and continue it to the horizon on which the vanishing point lies.

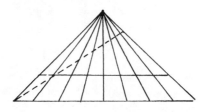

Where this diagonal crosses the other perspective lines shows where to draw the rest of the horizontals:

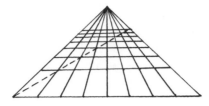

This was Alberti's "legitimate construction" (*costruzione legittima*). It was neither the first nor the last time that the asymmetry of a diagonal would win the day.

All through the Renaissance, artistic practice begot a flurry of mathematical insights to support and extend it, but not until the early years of the nineteenth century was an organic geometry developed which added as many vanishing points to the Euclidean plane as there are directions, and a circumscribing horizon as well, a "line at infinity" for all those "points at infinity" to lie on. What the eye proposes, Mind disposes—but it is always Mind incarnated in some particular mind.

The mind in this case belonged to a young French lieutenant of engineering in Napoleon's army. Jean-Victor Poncelet was a man of extraordinary willpower and character (at fifteen he had trained his dog to wake him at dawn so he could get back to his studies; the dog often found him asleep at his desk). At twenty-four he marched into Russia with the Grande Armée and was left for dead at the battle of Krasnoi, near Smolensk, in November 1812. The soldiers of the victorious Field Marshal Prince Miliradovitch recognized his officer's insignia and carried him off for interrogation, which saved him from death but condemned him to walking four months, and six hundred frozen miles, over the silent long plains to prison at Saratov, on the Volga. To keep up his spirits during the two years there, he tried to remember the mathematics he had studied, in a different life, just a few years before, at the Ecole Polytechnique in Paris. But the spiny demonstrations, the abstractions and generalizations, had perished with his comrades in the cold.

He began to build mathematics up from fundamentals again, trading his scanty rations for paper, making his own ink, and using the walls of his cell as a blackboard. Soon he

Poncelet (1788–1867). Loyal to his youth, he published in age his early work unedited by hindsight; loyal to France, he wasted his geometric foresight on its bureaucracy.

found his mind moving over vaster plains than those of Russia, and beyond the geometry he had been taught. "Oh God!" said Hamlet, "I could be bounded in a nutshell and count myself a king of infinite space." Perhaps being so bound in Saratov was what made Poncelet the king of projective geometry.

What sort of geometry can this be, where parallels meet? How can we picture, or even conceive of, a plane on which Alberti's *horizontals* also meet at a vanishing point—where no matter which way you look parallel lines converge, so that in fact there are no parallels at all? Isn't such an idea repellent to thought and repugnant to the world?

It certainly was to the world of Euclidean geometry. Some of the best mathematicians had tried for two millennia to prove what *must* be more than a mere postulate: that on a plane there is one and only one line, m, parallel to another line, ℓ, through a point P not on ℓ.

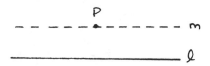

Gerolamo Saccheri (1667–1733) spent years trying to vindicate Euclid and, ironically, developed without realizing it most of the ideas of a geometry with *many* parallels to a given line though a point P not on it. Johann Heinrich Lambert, a generation later, tried to solve the parallel problem by looking at polygons on an unimaginable sphere of imaginary radius. The failure of all these attempts led even Gauss to speak of the parallel issue as the shameful part of mathematics, and to suspect, as did others, that if the existence and uniqueness of parallels was merely postulated, the opposite could be postulated as readily. There was, besides, a certain irritating asymmetry to Euclidean geometry: some lines had a point in common, others had none:

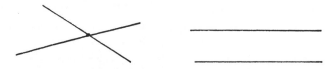

Why not restore symmetry by adding in the missing points: for all lines on the plane parallel to one another in a fixed direction, add just one point "at infinity" where they all meet:

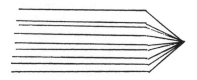

(the shape resulting from trying to picture this may have led to the bundle of lines being called a *pencil*).* We don't add two points at infinity for a pencil—one left, one right, or one west, one east—in order to preserve the postulate that two lines can't meet more than once. Along with these special points, add in the special line we spoke of on page 204: the line at infinity on which, like an ultimate horizon, these special points glitter. This completes the Euclidean to the *Projective* plane, which you might try to picture like this:

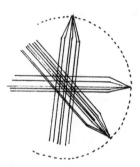

The pencils, swung through 180°, trace out the curve of the far horizon.

After mathematicians had spent a long time looking at it this way and that, the Projective plane turned out to be much simpler than Euclid's, with a packet of axioms even smaller than it seems:

P1: Any two points lie on exactly one line.
P2: Any two lines meet at exactly one point.
P3: There are (at least) three non-collinear points.
P4: At least three points lie on every line.

(Those last two axioms are to satisfy the inner Hilbert: "Does it exist?")

How can we look on the plane these axioms create and see it as it really is, without having to peer through a veil, or put up with such distortions as those playful "pencils"? You can no more expect to invite the infinite into your cozy world with impunity than hope that Alcibiades won't carry off half the silver from your feast. What we can do, however, is incarnate the projective plane in different ways, and by savoring the oddities of each, come better to appreciate its character.

There are several models of the four axioms: here is a surprising one. The objects themselves aren't surprising: the points are the familiar dots and the lines the conventional streaks—but there will turn out to be

---

*How far toward pure formalism are you willing to go? Would you agree to having the "point" at infinity added to this pencil of lines be *nothing other than that pencil itself?*

very few of each. Start with the three non-collinear points that the third axiom demands—call them A, B, and C:

A •

B •   • C

Now to fulfill P1 make lines through each pair of points,

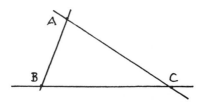

and in a dream of Euclid construct another line through A, as if it were to be parallel to BC:

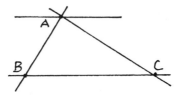

But P2 says it can't be: the new line must meet line BC in a new point D:

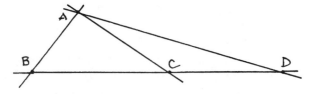

What begins as a parallel to AB through C must meet it in a point E, intersecting AD at F along the way

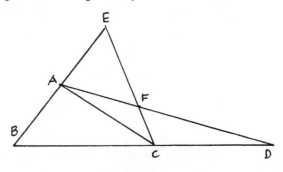

and satisfying P1 again with a line through B and F gives us a new point
G where this line meets AC:

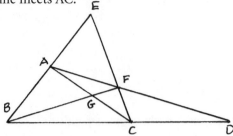

You probably think that this process of adding new points at the urging
of P2 and new lines at the behest of P1 will go on forever, generating a
model with an infinite number of points and lines—but in fact we have
all the points we need and all but one of the lines. D and E need to be
collinear, as do E and G—and so do D and G. Why not satisfy all three
demands at once with the drunken "line" DEG?

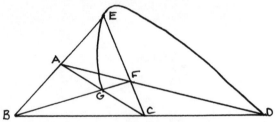

If you object that DEG is no line at all but a wandering path, remember
that "line" is an undefined term: only custom (and Euclidean custom at
that) asks that lines be straight. What matters here is simply—and startlingly—this: our model with its seven points and seven lines fulfills the
four axioms of the projective plane (we met the requirements of P4 without even having to think about them).

This model may satisfy the axioms but it hardly satisfies the mind.
Weren't we supposed to acknowledge that if it wasn't beautiful it wasn't
mathematics? Very well. Recall that in Chapter Five we found the incenter
of any triangle: the point where the angle bisectors meet, which is the
center of the "incircle" tangent to the triangle's three sides:

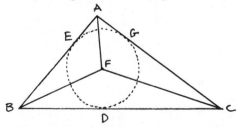

Can't we now freshly see this triangle with its angle bisectors, incenter,
and inscribed circle as the seven point and seven line projective plane?

True, one of the "lines" *looks like* a circle, but that is only because the doors of our projective perception have not yet been thoroughly cleansed.

How could such a cramped figure embody a geometry of spatial infinity? Other models are slightly larger, but finite too. We can construct one with thirteen points and thirteen lines, but if we start as we just did, putting down dots and running streaks through them, we will quickly get into a tangle resembling nothing so much as the web of a spider on LSD. The visual has always helped us—but our stubbornly Euclidean intuition means that it hinders us here. On the premise that what the eye can't see, the heart won't grieve for, let's turn momentarily away from sight altogether and think of our points as letters and the lines as their combinations. We will number these lines, only asking that each have exactly four "points" on it (that is, four different letters in its set).

$$\text{Line 1} = \{a, b, c, d\}\ .$$

Now P3 requires that there be a point—a letter—not in this set: make it the letter e. Then we must have a new line with a and e on (or in) it: P1 tells us that none of b, c, or d can be on this line as well—so we need two fresh points, f and g:

$$\text{Line 2} = \{a, e, f, g\}\ .$$

We will go on systematically in this way, making sure that any two letters lie in a unique set and that each set contains four letters. The fear that the bookkeeping will lead us to infinite excess is gradually put to rest as the combinations both needed and possible converge:

```
Line 1  = {a b  c  d}
Line 2  = {a        e  f  g}
Line 3  = {  b      e        h  i}
Line 4  = {  b         f           j  k}
Line 5  = {  b            g              l  m}
Line 6  = {     c   e                 j  l}
Line 7  = {     c      f  h                 m}
Line 8  = {     c         g  i        k}
Line 9  = {        d  e               k  m}
Line 10 = {        d      f  i              l}
Line 11 = {        d         g  h  j}
Line 12 = {a                    h     k  l}
Line 13 = {a                       i  j        m}
```

These letters and sets of letters obey each of our four axioms and so constitute a model of the projective plane.

Such a combinatorial exercise may lead you to agree with the nineteenth-century mathematician J. J. Sylvester: "Brindley, the engineer, once said that rivers were made to feed navigable canals; I feel almost tempted to say that space was made for feeding mathematical invention." Or it may cause you suddenly to reconsider the projective plane: it isn't a kind of space after all. It is a structure, a system of relations, which we could, if we chose, embody in space—but it is no more native to space than is the transmigrating soul to a particular creature's body. Must this then not be true of Euclidean "space" as well, or of anything generated by a collection of axioms?

We could go on to accountants' heaven with projective planes having 21 points and 21 lines, each with 5 points, or 31 of each (6 points to a line) or 43, 57, 73—in fact $n^2 - n + 1$ for any natural number n from 3 on, with n points on each of those lines—and so create an infinite number of finite models of the projective plane! But to nourish our starving intuition, let's look at one last visual model of this geometry, as wildly different from any of these as each is from its siblings: the thistle.

Picture the thistle's spines radiating out from a common core in every direction—or if that is too prickly, turn it into a Kooshball, but with infinitely many rubber threads rather than a mere 5000. The spines or threads may be as long as you choose—infinitely long, if you wish. You probably think that these will be the lines of our projective plane—but the surprise is this: they represent the *points*. Recall once more that "point" and "line" are undefined terms, so we may model them as perversely as we will, if only they behave according to the four axioms.

What then will stand for lines? Any two of these spines intersect at the center:

and back in the bucolic days of Euclidean geometry, two intersecting lines defined a plane. Each such plane will act as a *line* here. This makes sense: if our points look like lines, our lines must look like planes.

We now have to check: do any two points lie on exactly one line? That is, do any two spines or threads lie on a distinct plane? Yes, as you saw above, or as reinterpreted here:

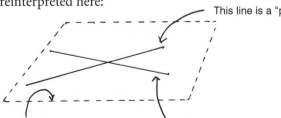

This line is a "point".

This plane is the "line" they lie on.   This line is another "point".

Do two lines meet in exactly one point? Our translator interprets: do two planes meet in exactly one line through the center? Again, yes:

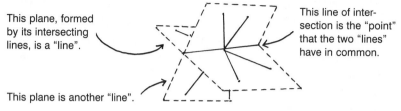

This plane, formed by its intersecting lines, is a "line".

This line of intersection is the "point" that the two "lines" have in common.

This plane is another "line".

Are there three non-collinear points? That is, are there three threads of the Kooshball that aren't all on the same plane? Here is an example:

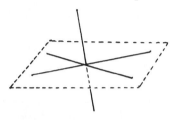

Three concurrent lines not all in the same plane—i.e., three "points" not all on the same "line".

And finally, has every line (that is, plane) at least three points (i.e., lines through the center) on it? Of course:

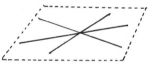

"Koosh" may be the sound that it makes when it lands in your hand—but what the Kooshball tells us is that we need three Euclidean dimensions to represent two of projective space; and that a model as far-fetched as this captures the structure latent in those four axioms as fully as does the seven- or thirteen-point plane, or the Euclidean plane completed with points and the line they lie on at infinity. When next you see the soft explosion of chrysanthemum fireworks in the summer night, or pick a humble burr off the hem of your coat, remember the projective plane.

If you are tempted to ask about any of these models: "Which is that special line at infinity in it, and which the special points?" we return the question to you with interest. Go back to our first model on page 206 (though it deserves a more dignified name than that, being no mere example but a very exemplar): the completed Euclidean plane. *After* it was completed—once any two lines met in a point and any two points lay on a line—could we really pick out the points or the line at infinity? The projective axioms have homogenized everything: these are all just points, just lines, obeying four laws. The desire to bring the infinite into our garden has had the unexpected consequence of giving all our plants double names. As we trim and tend the growths and watch patterns emerging among their patterns, novelties will merge into a new familiarity that satisfies desires we don't yet know we have. It isn't that we get what we want, as Proust once remarked, but that we come to want what we get.

We begin to acclimate ourselves to this landscape by first observing that there must be three non-concurrent lines in it: for the three non-collinear points that axiom P3 gave us will have lines through each pair of them by P1;

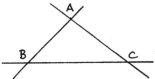

these can't *all* concur if P1 is still to hold. This is one of those truths you may think too trivial to mention, but we will soon profit from it in an unexpected way.

Slightly less obvious is a second observation of the same figure: at least three lines must meet at every point, for there will always be two points, such as A and B, which aren't both collinear with C (the point in question), and given the usual three lines through the pairs,

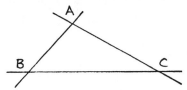

we will get two meeting at C. But P4 gives us another point D somewhere on AB, and DC is the third line going through C.

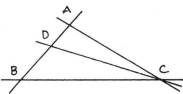

Having warmed up with these two exercises, look again at that packet of four axioms we said might be even smaller than it seemed. P1 and P2 oddly echo each other:

P1: Any two points   lie on   exactly one line.
P2: Any two lines   meet at   exactly one point.

Our two observations now allow us to echo P3 and P4:

P3: There are at least three   non-collinear points.
Observation One: There are at least three   non-concurrent lines.

P4: At least three points lie on every line.
Observation Two: At least three lines meet at every point.

Step back for a moment, as geometers after Poncelet finally managed to do, in order to view these paired statements from the right vantage point. What we see is something uncanny. Take any true statement in this projective geometry and exchange the nouns "point" and "line" for one another wherever they occur, and also their appropriate verbs, "lie on" and "meet at". Call this new statement the *dual* of the first. It will clearly be a different, but equally true, statement!

Why? Because a statement is true if it follows from the axioms—but as we have seen, the dual of each axiom is either another axiom or (as with the two observations) follows immediately from the axioms. This means that any theorem about some configuration of lines and points will yield another theorem with an identical structure about points and lines! Or to put it with disturbing vividness: if Euclidean custom leads you to picture your points like this: • and your lines like this: ⸻, well and good; but if, in projective geometry, you choose to draw your points thus: ⸺ and your lines so: •, nothing will be amiss. We found that we couldn't tell finite and infinite apart—now we can't even make out what are lines, what points! A mathematician named C. J. Keyser wrote in 1908: "Projective Geometry: a boundless domain of countless fields where reals and imaginaries, finites and infinites, enter on equal terms, where the spirit delights in the artistic balance and symmetric interplay of a kind of conceptual and logical counterpoint—an enchanted realm where thought is double and flows throughout in parallel streams."

Rather than sharing Keyser's enthusiasm, you may feel the sort of queasiness that comes with the first imperceptible tremors of an earthquake. We need to bring some sort of order by focusing on the *core* of this geometry: perspective and projection. In Euclidean geometry similarity and congruence were the key relations among triangles. Let's see how two triangles are most naturally related here. Alberti's Veil gives us the answer at once:

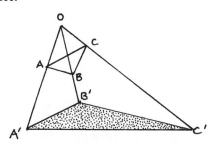

If you look at △ABC from a point of view O, △A′B′C′ is its image: or think of O as a light; then △A′B′C′ is its shadow. △A′B′C′ is *in perspective* with △ABC. Of course in this land of doubles, △ABC is just as much the image or shadow of △A′B′C′—but that's all right: the two triangles are *in perspective* when viewed from O, their *center of perspectivity* (just as on the Euclidean plane, the relations of similarity and congruence are symmetrical). Let's write:

$$\triangle ABC \overset{\scriptscriptstyle O}{\overline{\wedge}} \triangle A'B'C'$$

to mean that the two are perspective from O; that is, the paired vertices are lined up on rays from O: O, A, A′ are collinear, as are O, B, B′, and O, C, C′.

We might even do one perspectivity after another:

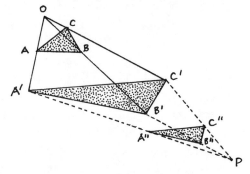

Here △ABC $\overset{\scriptscriptstyle O}{\overline{\wedge}}$ △A′B′C′, but also △A′B′C′ $\overset{\scriptscriptstyle P}{\overline{\wedge}}$ △A″B″C″. This chain of two perspectivities (from different points of view) we'll call a *projection*, and say we have projected △ABC onto △A″B″C″ (or vice versa) via this chain. A projection can have as many links as you choose—and we'll grant the title "projection" even to the single link of one perspectivity.

Where has this gotten us? Aren't things worse than ever? Two triangles in perspective certainly needn't be congruent—nor even similar; they probably haven't the same area and one triangle might even be acute and the other obtuse!

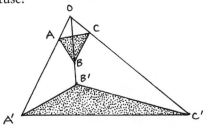

With such a feeble relation between them, how could we hope to have anything as interesting as the collinearities and concurrencies of Chapter Five? What shall abide the coming of projection? Is not all changed in

the twinkling of an eye? We relied on congruence in geometry and equality in algebra in order to transform one thing into another and see what
nevertheless remained invariant; yet here all is seeming and shadow, with
no objective form.*

Let the light of the golden seventeenth century organize these seemings
into sense. A self-taught French architect and engineer, Girard Desargues,
discovered a new and even more profound invariant of the projective plane.

He leads us to look once more at the simple, defining situation in this
geometry: two triangles in a perspective drawing on a plane:

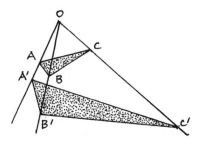

But you are seeing them only in part, he says: line segments, not lines.
Extend, for example, sides AB and A′B′ until they meet (as they must) at
some point P.

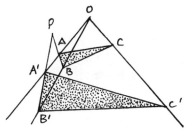

We are in danger of cluttering up the picture with too many lines—but
go on, he says, and find where the other paired sides meet at Q and R:

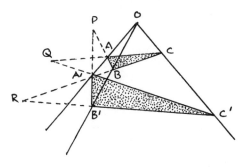

---

*If you worry about how much things seem to be slatting around on the projective plane, a theo-
rem lurks in the Appendix through which they are miraculously made fast.

We have been in a situation like this before: those three centers of a triangle, in Chapter Five, that *had* to be collinear. Is it an accident that P, Q, and R seem to be collinear too?*

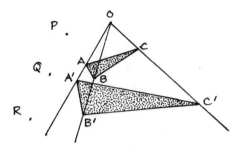

To show that in fact they *must* be, we travel, as always, elsewhere: not back to Euclid now but ahead into three dimensional projective space, P^3, whose six axioms—aimed at preventing parallels—are as straightforward as those of the projective plane. They include such assertions as that a line must intersect any plane in a point, and any two planes must meet in a line, and that there must be four non-coplanar points. Here is Desargues's gem of a proof that the paired sides of two triangles, perspective from a point, meet in three points that are collinear.

If the two triangles ΔABC and ΔA′B′C′ lie on *different* planes, N and M, and are perspective from some point O on neither plane, then their paired sides when extended must meet in three points that lie on the line ℓ, where the planes M and N intersect.

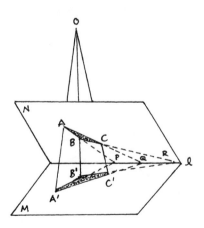

---

*We needn't have pictured the line that these three points lie on as "straight" but do so to accommodate our Euclidean vision. That what is to come also works on the Euclidean plane shows that once more we are entertaining a visitor there who has traveled from his projective home.

Why? Because lines OAA′ and OBB′, for example, intersect (at O) and hence form a plane—call it T. AB and A′B′ are two lines on this plane and so must intersect at a point—call it P. Since AB is on N and A′B′ on M, P is on each of *these* planes and so must lie on their intersection, the line ℓ, which is the hinge between the two planes. The same argument works for Q and R, so that all three lie on ℓ.

This is all very well, but not quite what we wanted. We need to deduce the same result when ΔABC and ΔA′B′C′ are on the *same* plane. Here's Desargues's architectural masterpiece.

We have ΔABC and ΔA′B′C′ on one plane—call it V—and perspective from a point O on this plane.

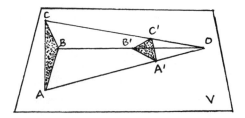

We know (from the axioms for projective space) that there is a point S not on the plane, so consider the line on which S and O lie (any two points lie on a line). Every line in projective geometry has at least three points, so there is another point—call it S′—on this line.

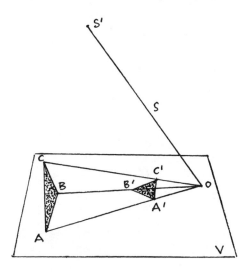

Now we will simply build Alberti's visual pyramids. Construct lines of sight from S to A, B, and C, and from S′ to A′, B′, and C′.

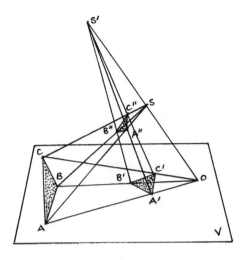

What may look confusing is really two pyramids intersecting, since SA, for example, and S′A′ must meet at some point—call it A″. Why must they meet? Because lines OSS′ and OAA′ meet at O, and once more, two intersecting lines form a plane. SA and S′A′ are lines on this plane, so must intersect.

Again the same kind of thinking shows us that SB and S′B′ intersect (at B″) and SC and S′C′ (at C″). A″, B″, and C″ are the vertices of that small triangle floating above plane V—the intersection of the two pyramids from S and S′.*

Now, with Desargues's eye, look steadily at what he has built and remember the fundamental power of transitive thinking. The floating triangle ΔA″B″C″ and ΔABC are on different planes but perspective from point S—hence, by Desargues's proof for triangles on *different* planes, their paired sides, when extended, meet at three points on one line: the line ℓ where plane V intersects the plane (which we haven't drawn in) of ΔA″B″C″. Call those points P, Q, and R.

The floating triangle ΔA″B″C″ and ΔA′B′C′ are also on different planes, but perspective from point S′—hence, again their paired sides, when extended, meet at three points on line ℓ. *These must be the same three points*, since A″B″, for example, intersects ℓ with AB at P and intersects it again with A′B′—but one line cannot intersect another in more than a single point.

By going up into a third dimension and returning, Desargues has shown that two *coplanar* triangles, perspective from a point, are also

---

*If A″, B″, and C″ were collinear, A, B, and C would be too—and we began with them forming a triangle.

"perspective from a line" (a condensed way of phrasing his conclusion). This line on which the paired sides meet is called the *axis of perspectivity*. We can relish his insight now as if it lay wholly on the plane.

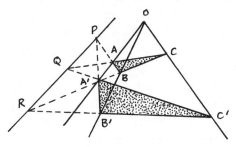

This being projective geometry, we are enticed into looking at Desargues's configuration from several different points of view. The first is duality. Since we now have the theorem: "If two triangles are perspective from a point, then they are perspective from a line," its dual must also be true: "If two triangles are perspective from a line, then they are perspective from a point."

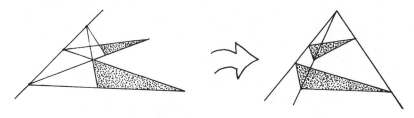

This dual is by no means as obvious as the original statement, but its proof grows beside the double river that waters this land.

At least as remarkable is the following exercise in looking at things askew. We will draw once more the "Desargues configuration" of ten points and ten lines. ΔABC and ΔA'B'C' are perspective from O and hence from line PQR. Now blink, and settle your seeing on any point other than O: choose, for example, C, and call *it* the center of perspectivity. Look—a new planet swims into our ken: ΔOAB and ΔQRC' are perspective from point C, and also from line A'B'P!

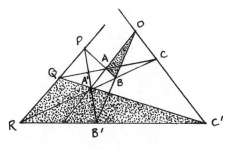

Choose another one of the ten points in this configuration as the center of perspectivity: again two triangles line up with it, and a fresh axis of perspectivity. Just as no point turned out to be a special "point at infinity", so none is a special "center of perspectivity". There are *ten* distinct "Desarguean configurations" compiled in this one—more collinearities and concurrencies than in all of Chapter Five, more ambiguities than in the most hypermodern novel.

Perhaps the most disconcerting reflection is this. We proved Desargues's theorem about the projective plane by moving into projective three-space. We had to: there can be no proof of it confined to the plane itself, making this particular fetching from afar not a *jeu d'esprit* but a necessity. People therefore tend to speak not of Desargues's theorem but Desargues's "theorem", since it is a theorem (as is its dual) only for projective planes when they are thought of as part of projective three-space. For an arbitrary projective plane, not similarly ensconced, his "theorem" is only an axiom—whose contrary is as easily affirmed (though at first perhaps not as cordially deemed worthy of belief). It is as if Desargues's conclusion were the shadow cast on the plane by a proof elsewhere. The union which set out against Euclid has loosened into a confederation of projective geometries.

∞

Projective planes in projective space—planes on which Desargues's theorem holds—are so rich that we can never gather up all their treasures. In this atmosphere thick with duality, it will come as no surprise to find that what were ends soon turn into means. Take, for example, the theorem in Chapter Five for which we had a whole volley of proofs: the medians of a triangle are concurrent. Let us bring yet one more proof—perhaps the most beautiful—from the distant projective plane.

Instead of drawing in any of the medians (so artful is this proof), let's just mark the midpoints D, E, and F of ΔABC's three sides:

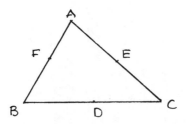

the line joining the midpoints we know (by similar triangles) is parallel to the base: so FE ∥ BC, FD ∥ AC, DE ∥ AB:

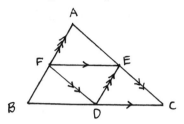

But parallel lines meet on the line at infinity—that is, FE coincides with BC at some point P there, FD with AC at a Q, DE with AB at some R. In other words, triangles ABC and DEF are perspective from a line (at infinity though it may be). Hence by the dual of Desargues's "theorem", these two triangles are perspective from a point—that is, there is an O at which AD, BE, and CF are concurrent—as we wished to show. We are looking straight down on Alberti's visual pyramid.

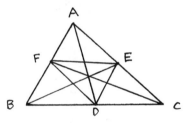

As J.B.S. Haldane once almost said: Mathematics is not only queerer than we suppose, but queerer than we *can* suppose.

∞

We will end this chapter on the endless with a magic trick. The best of these give the audience so much freedom to choose that you can't believe they could ever work—or if they do, it must be because of hidden accomplices. We love our freedom until it verges on an almost synonymous lawlessness at one extreme, a hint of subversive powers at the other.

So pick a line, any line, as the card sharpers say—and then pick another.

Next, choose any three points you like on the first, and any three on the second.

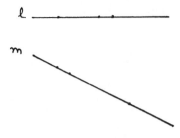

Please label the points on the first line A, B, and C—but again, in any order you choose; and (in any order), A′, B′, C′ on the second line.

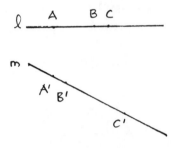

Now (while our assistant dusts off the vanishing points) connect A to B′ and B to A′, and call P the point where AB′ and A′B cross (we are still on the projective plane, so these lines *will* cross).

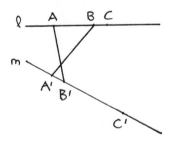

Let AC′ intersect A′C at Q, and BC′ meet B′C at R.

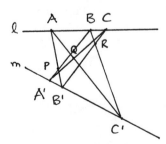

Abracadabra! P, Q, and R will be collinear. Should you care to redraw or relabel to see if this still looks true, we will entertain you the while with Hilbert's remark that a mathematical problem should be clear and easy to understand, since complication is abhorrent; should be difficult enough to entice us but not completely inaccessible ("lest it mock our efforts"); and should be significant: "a guidepost on the tortuous path to hidden truths."

Once you have convinced yourself experimentally that our claim is just, we can indulge in the different sort of conviction that comes from a proof—and its very different sort of pleasure as well: experiments generate wonder; proofs conclude with awe. Let's begin by adding to our diagram the point O where lines ℓ and m meet. We will draw the line PQ and prove that R is on it.

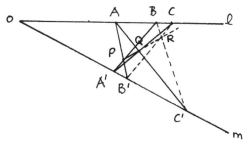

This is where the magician lets out the rabbit that was all the time in his hat: take that line PQ as the line at infinity. What have we just done, and who said we could do it? Remember that once the Euclidean plane is completed by adding to it (along with special points) the line at infinity, all lines look and behave alike, so *any one* can now be rechristened the line at infinity! This move is like a modulation in a late Beethoven quartet: inspired, outrageous, transforming. It trumps the original freedom of choice with a freedom of its own.

Since P is now the point at infinity where AB′ and A′B meet, they are in the old Euclidean sense parallel; as are AC′ and A′C, since they meet at Q on the line at infinity. If you like, you may think of what we've done this way: we have taken advantage of being on the projective plane by choosing our point of view so that these pairs of lines are parallel. Our configuration would now look like this:

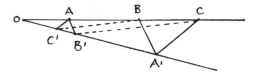

To prove that R (where BC′ and B′C meet) is on this line now amounts to proving BC′ ∥ B′C.

Let's assign lengths q, s, t, u, v, and w to segments in the diagram as follows:

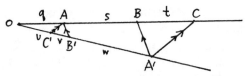

Since $\triangle OAB' \sim \triangle OBA'$, the sides are in proportion: that is,

$$\frac{q}{u+v} = \frac{q+s}{(u+v+w)} \, ,$$

so

$$\frac{q(u+v+w)}{u+v} = q+s \, ,$$

or

$$q+s = \frac{q(u+v+w)}{u+v} \, .$$

And since $\triangle OAC' \sim \triangle OCA'$, $\dfrac{q}{u} = \dfrac{(q+s+t)}{(u+v+w)}$, so

$$q = \frac{u(q+s+t)}{(u+v+w)}$$

and

$$\frac{q(u+v+w)}{q+s+t} = u \, ,$$

or

$$u = \frac{q(u+v+w)}{q+s+t} \, .$$

Hence we can write:

$$\frac{(q+s)}{u} = \frac{\dfrac{q(u+v+w)}{u+v}}{\dfrac{q(u+v+w)}{q+s+t}} \, .$$

Let's simplify this ungainly double-decker by dividing both its numerator and denominator by q(u + v + w). We will then have

$$\frac{(q+s)}{u} = \frac{\dfrac{1}{u+v}}{\dfrac{1}{q+s+t}} = \frac{q+s+t}{u+v} \ .$$

But this implies that $\Delta OBC' \sim \Delta OCB'$, so that $C'B \parallel B'C$, and their meeting point, R, is indeed collinear with P and Q on the line at infinity.

It would be nice to end with a fanfare—an illustration of these similar triangles revealing the collinearity of P, Q, and R. But notice that AB' and $A'B$ must be simultaneously parallel and convergent at P. Since P is the point at infinity, the mind can see it—but the hand trembles too much to make a drawing.

What we have just done is as neat a piece of cross-ruffing as you are likely to see (if you object to such playing off of one kind of plane against another, a different proof of this theorem, wholly in the projective idiom, is in the on-line Annex).

Which was more magical, the theorem or its proof? In either case the show isn't over. This theorem was first discovered by the witty Alexandrian geometer, Pappus, whom you may already have met in the appendix to Chapter Five (page 282): the man who proved the base angles of an isosceles triangle congruent by thinking of the triangle as congruent to its mirror image. A thousand years later, Pascal discovered that if you sprinkle these two triples of points anywhere on a circle's circumference, the same result holds:

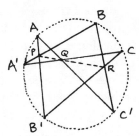

P, Q, and R are still collinear, no matter where you put them or how you label them.

Shall we push incredulity further toward the brink? Distort that circle into any sort of ellipse and P, Q, and R remain stubbornly perched on a single line:

225

What about a parabola?

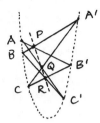

Dare we go to the extreme of a hyperbola's two branches?

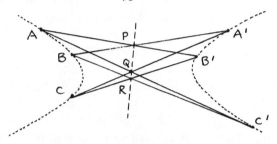

Could we risk even thinking about the duals of each of these theorems?

Hard to swallow as the proof of Pappus's theorem was with a pair of straight lines, won't trying to prove it in any of these four new configurations mock our efforts? Not at all. They will now be simplicity itself and a guidepost on the tortuous path to hidden truths (indeed, Hilbert called Pappus's theorem the most important in all of plane geometry, because Desargues's theorem, or any theorem about lines meeting on the plane, can be derived from it). The simplicity comes from noticing that a pair of lines, a circle, an ellipse, a parabola, and a hyperbola are all *conic sections*: slices, not through Alberti's visual pyramid, but through a palpable cone. They are projective transformations, therefore, of one another, when seen from the cone's apex (the hyperbola's second branch lies up in the cone's mirror image: extending, as always, the known into the new). A projective invariant of one will be invariant for all.

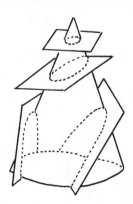

Projective geometry—so sprightly in its approaches, so profound in its results—is the contemplation of permanence behind change, animating the sculptural beauty of Euclid in a world of transformations. Peacock fumbled at this with his Principle of Permanence. The new and deeper sense it makes here was expressed as a *Principle of Continuity* by Poncelet in 1822: "If one figure is derived from another by a continuous change and the latter is as general as the former, then any property of the first figure can be asserted at once for the second." This principle, and the subsequent algebraic approach to projective geometry, underlie our modern facility for moving from one coordinate system to another— a facility that underlies, for example, Einstein's understanding of measurements made by observers differently situated and in motion relative to one another.

The Abyss

If you stare too long into an abyss, the abyss will stare back into you.

—Nietzsche

We would now spiral back to the counting with which the art of the infinite began, were it not that a little cloud on the mind's horizon—no bigger, really, than a man's hand—has from time to time troubled people telling their numbers. In the ninth century the star worshipper Thabit ibn Qurra, from Harran, argued that infinity could be split into two, three, or any number of parts, each of which would then have the same size as the whole: there is, for example, an infinite number of evens, but also of odds, so each half of infinity is infinity—and so on.

In 1638 Galileo argued that "equal", "greater", and "less" can't apply to infinite quantities because a line segment contains an infinity of points, so a longer segment would have to contain more than that infinity, which is impossible. And again, each natural number has its square matched with it:

$$
\begin{array}{ccc}
1 & \leftrightarrow & 1 \\
2 & \leftrightarrow & 4 \\
3 & \leftrightarrow & 9 \\
4 & \leftrightarrow & 16 \\
& \cdot & \\
& \cdot & \\
& \cdot & \\
n & \leftrightarrow & n^2 \, .
\end{array}
$$

Hence there must be exactly as many square numbers as there are naturals. Clearly, however, there are fewer, since the squares grow ever sparser as you go along.

Putting the question aside seemed for a long time the best way of coping with it. Infinite collections of numbers were too slippery to try comparing, as if we were being warded off from these higher mysteries by their power to confuse. We could speak negatively of approaching but never reaching a limit, or of there not being a last prime; and positively about what must be true for any (or was it every?) number—we could even make finite models of infinite planes. But how could a mind tucked into a little skull possibly grasp the infinite *as such*, or count its way through infinite multitudes?

The work of a single man utterly changed our glib know-nothingness forever. What was to follow would be colored by the strengths and weaknesses of his particular personality, rather than having the impersonal air we tend to associate with mathematics and collective work.

Georg Cantor must have been born in the imperative mood. This mattered at least as much as the intellectual climate of Germany in 1845. He was propelled through his youth by a torrent of a father ("Now my dear son! Believe me—to prevent the slander of open or secret enemies you need to acquire the greatest amount possible of the most basic knowledge. Whatever one neglects through premature extravagance is irretrievably lost, like lost innocence. . . . Your father, your family, have their eyes on *you*. . ."). Even more compelling was what he described throughout his life as a secret voice—within, above, unknown—a "more powerful energy" that spoke through him. He always looked for the face behind the mask and then for the mask behind that. Docile at home, domineering among colleagues, playful in mathematics and humorless in his wrangles with mathematicians, he was as close to a reincarnation of Alcibiades as nineteenth-century Germany could produce—not only in his enthusiastic energy and wild daring, but in the ferocious way he fought when cornered—Alcibiades by Phrygians, Cantor by ideas.

Cantor as a young man.

Let's return, with Cantor's inflexible will and malleable imagination speaking within us, to Galileo's problem of the natural numbers and their squares. Since each number has its unique square and each square corresponds to a single natural, it certainly seems right that there are just as many squares as there are naturals, for all that the squares are *scattered* among them. Let skepticism give way to astonishment and astonishment to experimental candor: let's follow where this observation leads.

Other such correspondences come trooping after. Although only every second natural number is even, there must nevertheless be exactly as many of them as naturals, since each natural is perfectly matched with its double:

$$
\begin{array}{ccc}
N & & 2N \\
1 & \leftrightarrow & 2 \\
2 & \leftrightarrow & 4 \\
3 & \leftrightarrow & 6 \\
4 & \leftrightarrow & 8 \\
\cdot & \cdot & \cdot \\
\cdot & \cdot & \cdot \\
n & \leftrightarrow & 2n
\end{array}
$$

To say these matchings-up show that there are as many of one kind as of the other needs, of course, a very bold leap of thought. We are taking each sort as a *completed whole: all* the naturals match up perfectly with *all* the evens, or with *all* the odds. Stop short and the correspondence breaks down: there are only 50 evens among the first 100 naturals, for example, and 50 odds.

The multiples of 3 are even thinner on the ground than those of 2— and yet once again, there are just as many of them as of the naturals they are selected from:

$$
\begin{array}{ccc}
N & & 3N \\
1 & \leftrightarrow & 3 \\
2 & \leftrightarrow & 6 \\
3 & \leftrightarrow & 9 \\
4 & \leftrightarrow & 12 \\
\cdot & \cdot & \cdot \\
\cdot & \cdot & \cdot \\
n & \leftrightarrow & 3n
\end{array}
$$

We could walk over N in seven league boots and take just as many paces as the numbers we stride through:

$$
\begin{array}{ccc}
N & & 7N \\
1 & \leftrightarrow & 7 \\
2 & \leftrightarrow & 14 \\
3 & \leftrightarrow & 21 \\
4 & \leftrightarrow & 28 \\
\cdot & \cdot & \cdot \\
\cdot & \cdot & \cdot \\
n & \leftrightarrow & 7n
\end{array}
$$

How little a step now for the mind to invoke its own sort of infinity and declare: for *any* natural number m, there are just as many multiples of m as there are natural numbers altogether:

$$
\begin{array}{ccc}
N & & mN \\
1 & \leftrightarrow & m \\
2 & \leftrightarrow & 2m \\
3 & \leftrightarrow & 3m \\
4 & \leftrightarrow & 4m \\
\cdot & \cdot & \cdot \\
\cdot & \cdot & \cdot \\
n & \leftrightarrow & nm
\end{array}
$$

If you ask: how many is that? we could answer in terms of *cardinal numbers*, which read off what the *size* of a set is—that is, how many elements (in whatever order) it contains. The set with a cabbage, a goose, and a fox:

$$\{cabbage, goose, fox\}$$

has cardinality three (and the problem is to keep it that way). So has the set

$$\{13, -8, 251\}.$$

The set with the first million counting numbers has cardinality one million. Here we could say: the sets N and mN have the *same* cardinality, as established by the one-to-one correspondence we made. Since we count by means of the natural numbers, we could also say that both sets are *countable*.

Had Cantor done nothing else, this insight would still have revolutionized our understanding of the infinite. What was a paradox could now be seen as a peculiar truth suggestive of truths perhaps yet more peculiar: the hallmark of mathematics at work. To say that Cantor did infinitely more would be an understatement. In Chapter One we remarked that matching up separate things with a sequence of numbers might seem of little consequence, but would take us beyond the moon. With the set of all the natural numbers (or all the multiples, if you wish, of 65,537), we are already well past it, yet hardly any distance along the path that Cantor took: as winding, as steep, as exhilarating as those he walked in the Harz mountains.

The crucial ideas in mathematics are always so simple as to seem intuitively clear: sets and making 1–1 correspondences between their members. You needn't even know how to count to do this, and the effect has always been spectacular. In the fourth century a nomadic army from the

East rode through the Caucasus into Armenia. "No one could number the vastness of the cavalry contingents," wrote a contemporary chronicler, "so every man was ordered to carry a stone so as to throw it down . . . so as to form a mound . . . a fearful sign left for the morrow to understand past events. And wherever they passed, they left such markers at every crossroad along their way." In Scotland the Cairn of Remembrance still stands at Invercauld, where the Farquharsons each set down a stone before battle—and those who survived took each his stone back home.

Although the word "set" first took on its technical meaning with Cantor, surely it stands for what we are all born knowing, as we observed in Chapter Two. Sords of mallards and prides of lions tickle our easy aptitude for making a many into a one. It is almost more comfortable to think, for example, of your old *gang* taking on a *bunch* of hoodlums than having to deal with single people who have faces and friends. If counting, as mathematicians know from bitter experience, is harder even than hitting a round ball with a round bat (which Ted Williams said is the hardest thing there is), certainly the young Cantor made it significantly easier by recognizing "set" as the central noun of the new language he was inventing. Its central verb was "to correspond". The correspondence between the members of the sets might be hard to find; the way you made it might seem artificial or bizarre—but once revealed, the two sets between which it ran had to have the same cardinality. Conversely, you must agree, if somehow you proved that no 1–1 correspondence could exist between the members of the two sets, then their cardinalities would have to be different. On such casual agreements momentous conclusions hang.

Let us continue, with Cantor, to learn again how to count—which may make us sympathize with birds and chimpanzees. Having found that any infinite sequence of the naturals is countable (not as great a surprise as it first seemed, if you think about it, since such a sequence will have a first term, then a second, third and so on—and ordering them thus in effect counts them), we are tempted to look in the opposite direction: not at subsets of N but at a set that includes it. This is the set Z of integers, with zero and the negatives of the naturals as well. Is it possible to put this set too in one-to-one correspondence with its subset N?

Yes, but with a slightly greater effort of the imagination. After starting with 0, just hop back and forth between the naturals and their negatives:

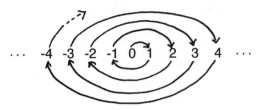

This establishes the match-ups:

N		Z
1	↔	0
2	↔	1
3	↔	−1
4	↔	2
5	↔	−2
6	↔	3
7	↔	−3
.	.	.
.	.	.

We have thus shown that the integers have the same cardinality as the naturals.

Such a clever way of pairing gives us the confidence now to think the unthinkable and face what Galileo shied away from: a more than infinity. For if you look at all the rational numbers Q, or even at just the positive rationals—the *set* of all these fractions—there are obviously more of them than there are natural numbers, since between any two fractions will lie another, until what was the space from one natural to the next will be crammed to bursting with them. You could look at it this way: any set as numerous as the naturals is countable, but how could you possibly count the positive rationals? How say, given one of them, which is the next, or even which is the first of them all? Between 0 and any candidate you name, another will crop up—and another . . .

If it is hard to conjure up the entrenched determination of Alcibiades, imagine Cantor prowling the margins of the forest of fractions, certain that they could be counted if only looked at from the right angle. So the chaos of an orchard seen from a passing train resolves itself for a moment into ordered rows.

Clearly their obvious order, from smaller to larger, doesn't help, because of the ceaseless in-betweens. They would have to be rearranged, like an orchard:

$$\frac{1}{1} \quad \frac{2}{1} \quad \frac{3}{1} \quad \frac{4}{1} \quad \frac{5}{1} \quad \frac{6}{1} \quad \frac{7}{1} \quad \frac{8}{1} \quad \frac{9}{1} \cdots$$

$$\frac{1}{2} \quad \frac{2}{2} \quad \frac{3}{2} \quad \frac{4}{2} \quad \frac{5}{2} \quad \frac{6}{2} \quad \frac{7}{2} \cdots$$

$$\frac{1}{3} \quad \frac{2}{3} \quad \frac{3}{3} \quad \frac{4}{3} \quad \frac{5}{3} \cdots$$

$$\cdot \quad \cdot \quad \cdot \quad \cdot \quad \cdot$$
$$\cdot \quad \cdot \quad \cdot \quad \cdot \quad \cdot$$

Rethinking the divisions on a line as this two-dimensional grid is the turn of thought we have met again and again: leaving the familiar and nearby to return, enlightened, from a distant land. Now at least we have what looks like a beginning: $\frac{1}{1}$ pinned in the upper left-hand corner. But next? to move steadily to the right, clicking off $\frac{2}{1}, \frac{3}{1}, \frac{4}{1}$ ... will count the top row but leave the vast acres of fractions beneath it untouched. To move steadily down from $\frac{1}{1}$ will number $\frac{1}{2}, \frac{1}{3}, \frac{1}{4}$, and all the Egyptian fractions at the expense of the endless columns to their right.

Let these two necessities mother invention by scratching the eternal itch of asymmetry. Go neither exclusively across nor exclusively down, but zigzag along diagonals through the planting:

$$
\begin{array}{ccccccccc}
\frac{1}{1} \rightarrow & \frac{2}{1} & \frac{3}{1} \rightarrow & \frac{4}{1} & \frac{5}{1} & \frac{6}{1} & \frac{7}{1} & \frac{8}{1} & \frac{9}{1} \cdots \\
& \swarrow & \nearrow & \swarrow & & & & & \\
\frac{1}{2} & \frac{2}{2} & \frac{3}{2} & \frac{4}{2} & \frac{5}{2} & \frac{6}{2} & \frac{7}{2} & \cdots & \\
\downarrow \nearrow & \swarrow & & & & & & & \\
\frac{1}{3} & \frac{2}{3} & \frac{3}{3} & \frac{4}{3} & \frac{5}{3} & \cdots & & & \\
\end{array}
$$

This is the path that will count the positive rationals, so long as it is walked judiciously. Each fraction must appear once—but only once—on our list, but the fifth entry here, $\frac{2}{2}$, is the first, $\frac{1}{1}$, in disguise. Very well: starting at $\frac{1}{1}$ follow this maze and count each entry in order, so long as it hasn't appeared before. Then we have a 1–1 correspondence between $\mathbb{N}$ and the positive rationals—$\mathbb{Q}^+$. It begins:

$$
\begin{array}{ccc}
\mathbb{N} & & \mathbb{Q}^+ \\
1 & \leftrightarrow & \frac{1}{1} \\
2 & \leftrightarrow & \frac{2}{1} \\
3 & \leftrightarrow & \frac{1}{2} \\
4 & \leftrightarrow & \frac{1}{3} \\
5 & \leftrightarrow & \frac{3}{1} \\
6 & \leftrightarrow & \frac{4}{1} \\
7 & \leftrightarrow & \frac{3}{2} \\
8 & \leftrightarrow & \frac{2}{3} \\
9 & \leftrightarrow & \frac{1}{4} \\
10 & \leftrightarrow & \frac{1}{5} \\
11 & \leftrightarrow & \frac{5}{1} \\
& \vdots &
\end{array}
$$

The eccentricity of this sequence makes sense only when you see from above the map of its two-dimensional source—but since each positive rational appears precisely once here, the sense beyond sense it conveys is that the set of positive rationals contains no larger an infinity than that of the natural numbers. "How many" must have nothing to do with "how dense".

Three thoughts come tumbling all in a rush. First, notice how the need for imagination has increased by quantum jumps through our three problems. To show that the squares or the evens or the multiples of any number m were countable took steadfast looking: letting the world instruct the eye. To count the integers we needed to free ourselves from thinking via succession so as to come up with the pert invention of hopping back and forth. To count the positive rationals we had to shake off linear thinking altogether and devise a two-step as precariously balanced as Harold Lloyd on an I-beam. The questions we ask beget means to answer them that grow past all expectation in refinement, and develop an arcana of their own.

Second, you now can see why we said, in Chapter Six, that if the bookkeeper in the brain really insisted on putting in order all those links in the infinite chains hanging down from the infinitely long chain of square root extension fields, he could do so: just diagonalize through them as Cantor inspired us to do.

Third, not just the positive but *all* the rationals are crying out for us to count them. It only takes combining our second and third techniques. Make the zigzag through the positive rationals and then make another, independently, through the negatives:

$$\frac{-5}{1} \quad \frac{-4}{1} \leftarrow \frac{-3}{1} \quad \frac{-2}{1} \leftarrow \frac{-1}{1} \qquad \frac{1}{1} \rightarrow \frac{2}{1} \quad \frac{3}{1} \rightarrow \frac{4}{1} \quad \frac{5}{1}$$

$$\frac{-5}{2} \quad \frac{-4}{2} \quad \frac{-3}{2} \quad \frac{-2}{2} \quad \frac{-1}{2} \qquad \frac{1}{2} \quad \frac{2}{2} \quad \frac{3}{2} \quad \frac{4}{2} \quad \frac{5}{2}$$

$$\frac{-3}{3} \quad \frac{-2}{3} \quad \frac{-1}{3} \qquad \frac{1}{3} \quad \frac{2}{3} \quad \frac{3}{3}$$

$$\frac{-2}{4} \quad \frac{-1}{4} \qquad \frac{1}{4} \quad \frac{2}{4}$$

$$\frac{-1}{5} \qquad \frac{1}{5}$$

We know we can put each of these sets, Q^+ and Q^-, in 1–1 correspondence with N. We also know that the set of evens, E, and the set of odds,

O, are in 1–1 correspondence with N. Transitivity and interleaving to the rescue: match the positive rationals, Q^+, with the even naturals, E, by way of that zigzag; and the negative rationals, Q^-, with the odds, O, in the same way. Then shuttling back and forth between odds and evens will put the totality of Q^+ and Q^- in 1–1 correspondence with N.

We have left out zero, and make good our omission by bumping the correspondence of Q^- with the odds over one, leaving the natural number 1 with no partner. Pair it up now with 0. This counts all of Q, as desired:

$$
\begin{array}{ccc}
N & & Q \\
1 & \leftrightarrow & 0 \\
2 & \leftrightarrow & \frac{1}{1} \\
3 & \leftrightarrow & \frac{-1}{1} \\
4 & \leftrightarrow & \frac{2}{1} \\
5 & \leftrightarrow & \frac{-2}{1} \\
6 & \leftrightarrow & \frac{1}{2} \\
7 & \leftrightarrow & \frac{-1}{2} \\
8 & \leftrightarrow & \frac{1}{3} \\
9 & \leftrightarrow & \frac{-1}{3} \\
10 & \leftrightarrow & \frac{3}{1} \\
11 & \leftrightarrow & \frac{-3}{1} \\
& \vdots &
\end{array}
$$

No matter how far-flung the rational you name, it will eventually put in an appearance on this list.

You will appreciate the exhilaration Cantor must have felt in winning such striking insights as this by going against the authority of the demigods Aristotle, Gauss, and his own contemporary Kronecker, who said that it was illegitimate to think of or deal with *completed* infinities. For them, as for almost all the world before Cantor, the infinite was *potential*. By making it actual he put infinite ensembles on a par with finite ones, rethinking "number" altogether in terms of "sets"—and so laid the foundations of modern mathematics. He also apparently put to rest the paradox people had somehow managed to live with, of seeing how absurd it must be to have a "more than infinity", yet being sure at the same time that there were more fractions than naturals. Keeping your thought compartmentalized helps to hold two such incompatibles in it, and will get you through many a difficult day.

Pick up your neo-Pythagorean talisman again to see what Cantor achieved. He matched up the members of the inner circle $\mathbb{N}$ with those of its surrounding $\mathbb{Z}$ and now with all the elements of the wider enclosure $\mathbb{Q}$, so that "infinity" had not many vague meanings but one, thanks to the notions "set" and "1–1 correspondence". $\mathbb{R}$ was the next candidate, with all of the rationals in it, but all the irrationals too. The irrationals: it was these, you recall, whose existence shattered the Pythagorean order of things. Once again we are about to fall off the edge of the world.

In 1872, while Cantor was on holiday in the Swiss village of Gersau—mountains before it, ravine behind—he chanced to meet Richard Dedekind. They recognized at once the affinity of their thought and continued the conversations begun there in an exchange of letters. On November 29, 1873, Cantor wrote to Dedekind that it seemed impossible to match up the naturals with the reals because the former were discrete and the latter made a continuum—". . . but nothing is gained by saying so, and as much as I incline to this opinion, I haven't been able to find the reason, which I keep working at; perhaps it is really very simple." On December 2, he added that he had been trying to deal with this for years and couldn't decide whether the difficulties were his or lay in the problem itself.

Then suddenly—on December 7—he wrote again: he had found a proof that the real numbers couldn't in any way be put into a 1–1 correspondence with the naturals—they were *uncountable*. Since the real numbers contain the naturals as a proper subset (every natural is a real, that is, but not every real is a natural), this suggests that in fact there are *more* of them: a larger infinity than that of $\mathbb{N}$ or the equinumerous $\mathbb{Z}$ and $\mathbb{Q}$.

You probably expect that so shattering a conclusion follows from a proof whose subtlety or abstruseness could never be contained in these pages—yet here it is, in a version Cantor came up with later: the most stunning work in the gallery of nineteenth-century art, and built once again on the strut of a slender diagonal.

Cantor had to show that there *was* no 1–1 correspondence between the sets $\mathbb{N}$ and $\mathbb{R}$—not just that he couldn't find one. The only logical approach to such negative statements was a proof by contradiction. His strategy would be to assume that a 1–1 correspondence *had* been made, and then to reduce this assumption, as they dismissively say, *ad absurdum*. His tactics involve first restricting his attention to a small subset of the reals: all those greater than 0 and less than 1. We write $(0, 1)$ to stand for this "open interval from 0 to 1", which you may picture as a segment of the real line without its end-points:

Any of these reals can of course be written as a decimal beginning "0.", and continuing with a string of digits. As we found in Chapter One, a pattern will emerge among those digits for any rational, but *not* for any irrational. In either sort, however, one of the ten digits from 0 to 9 will occupy each of its countably many decimal places. Cantor's proof turns on this banal observation.

For assume, now, that all of these decimals are in a 1–1 correspondence with the naturals, so that we can list them. Since the aim is to contradict this assumption, we can't specify how the listing is to be carried out: *any* possible way of arranging them must conclude against the same wall. We therefore need neutral markers to stand for whatever the digits in any entry may be. Since we are on better terms with subscripts after the adventures of Chapter Six, let a_{11} stand for the digit in the first decimal place of the first entry, a_{12} for the second digit there, and so on:

$$1 \leftrightarrow 0 . a_{11} a_{12} a_{13} a_{14} \cdots$$

The decimal matched up with 2 in our puzzling list will have entries a_{21}, a_{22}, and so on:

$$2 \leftrightarrow 0 . a_{21} a_{22} a_{23} a_{24} \cdots$$

so that the supposed one-to-one pairings-up of all the naturals with all the reals in (0, 1) will look like this:

$$1 \leftrightarrow 0 . a_{11} a_{12} a_{13} a_{14} \cdots$$
$$2 \leftrightarrow 0 . a_{21} a_{22} a_{23} a_{24} \cdots$$
$$3 \leftrightarrow 0 . a_{31} a_{32} a_{33} a_{34} \cdots$$
$$4 \leftrightarrow 0 . a_{41} a_{42} a_{43} a_{44} \cdots$$

.

.

Each entry continues forever (i.e., with as many decimal places as there are counting numbers), and there will be as many entries on the list as there are counting numbers.

We are supposing that this list is *complete*: every real in (0, 1) is somewhere on it, hence there are no forgotten or neglected real numbers in this interval that can be added on at the end—and a good thing too, since the list *has* no end. We are also supposing that no entry appears here twice: any two decimals listed must differ in *at least* one decimal place ($\frac{1}{9}$, for example, is listed somewhere: its decimal form is $0.\bar{1}$, that is, 0.11111... forever; and here too is the decimal with 1 in every decimal place—except for a 0 in the 93,247th place).

Now comes the diagonal stroke of genius. That first decimal place in the list's first entry, a_{11}, must, of course, be one of the digits from 0 to 9: for example, it is either 5, or not. Cantor asks us to create our own decimal number between 0 and 1 as follows. Like those on the list, it too will begin "0.", but its first decimal place will be determined by what a_{11} is. If a_{11} is 5, our number will have a 6; but if a_{11} isn't 5, we put a 5 in the first place of ours.

So far our decimal looks either like this: "0.6" or like this: "0.5".

To decide whether to put a 5 or a 6 in the second decimal place of ours, look at the second entry in the *second* decimal on the list: a_{22}. Again we act contrariwise: if a_{22} is 5 we will have 6 in our second decimal place; but if a_{22} isn't 5, 5 goes there in ours. We thus have 0.65, 0.66, 0.55, or 0.56, depending on what a_{11} and a_{22} were.

Continue filling the successive decimal places of our number with 5 or 6 in this mechanical way, looking at a_{33} for our third place, a_{44} for our fourth, and in general, sliding gracefully down this diagonal:

$$0 . a_{11} a_{12} a_{13} a_{14} a_{15} \cdots$$
$$0 . a_{21} a_{22} a_{23} a_{24} a_{25} \cdots$$
$$0 . a_{31} a_{32} a_{33} a_{34} a_{35} \cdots$$
$$0 . a_{41} a_{42} a_{43} a_{44} a_{45} \cdots$$
$$0 . a_{51} a_{52} a_{53} a_{54} a_{55} \cdots$$

whatever fills the nth decimal place of the nth entry, a_{nn}, determines whether we put 5 or 6 in the nth place of ours.

The real number we are building up has only 5s and 6s in its decimal places, and might begin like this:

$$0.55666565656656555 \ldots$$

Whatever it looks like, it is a perfectly good real number, somewhere to the right of center in (0, 1): more precisely, it will be between $\frac{5}{9} = .\overline{5}$ and $\frac{6}{9} = .\overline{6}$.

Notice, however, that it can't possibly be the first entry on the list, since it differs from that entry at least in the first decimal place. It can't be the second entry either, differing as it does from it in at least the second decimal place; nor the third, for the analogous reason, nor the

fourth—nor the nth. It cannot, in fact, be anywhere on this list that was supposed to contain *all* of the reals in (0,1) because it differs from every entry on it in at least one decimal place—and that is the contradiction.

This proof—as simple and subtle as all great art—throws open the gates to what Hilbert called Cantor's Paradise. If we can compare infinite cardinalities—if we understand the proof to show that there are *more* real numbers in (0,1) than there are naturals altogether—then we have just found a second and larger size of infinity (and the hairs on the back of the neck stand up at the hint of perhaps more). It is hard to think of a comparable shock to the life of the mind (unless it be the revelation that others think "I").

∞

Now we can return to Galileo's shorter and longer line segments. The open interval (0, 1) has more points on it than all the counting numbers in the world, although there is no end of *them*. What about the longer segment (0, 2)? An astonishingly simple proof—another "Look!"—shows that this longer segment contains just as many points as the first: there is a 1–1 correspondence between them.

Center the first segment above the second, and for the sake of the proof put on their missing end-points:

We know from the previous chapter what to do with these two lines: find their center of perspectivity, P:

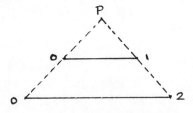

Now project the smaller interval onto the larger from P: each point in it matches up with a unique point on the other—and vice versa:

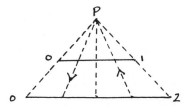

This projection establishes a 1–1 correspondence between them: the midpoint, $\frac{1}{2}$, of the first segment matches up with the midpoint, 1, of the second, and trigonometry will, if you want, give you the rest of the match-ups—but *de minimis non curat* Cantor. You may now eliminate the end points of each, but leave the P they created: it still shows that the cardinality of (0, 1) and of (0, 2) are the same.

Why stop here? Take some very short open interval, like (5, 5.1) and some very big one: (−3,000, 1,000,000). The same projection establishes the 1–1 correspondence between their elements:

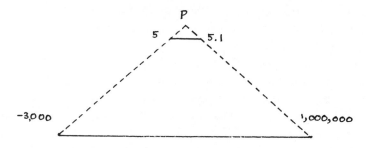

The number of points on the horizontal bar of the t in "horizontal" has exactly as many points on it as the line from the earth to Alpha Centauri. "How many" has nothing to do with "how long".

If space was created for feeding the imagination of geometers, counting was created for feeding Cantor's. The points on any line segment are gigantically equinumerous with those of any other—but what about the points on the entire real line, disappearing to negative and positive infinity? How can we show that the entire line has as many points altogether as even on the merest smudge of one of its segments—or that it has more? The proof by projection no longer works because the real line lacks end-points to pull up the sight-lines from. Here is ingenuity raised to the 13th power ("What lack we yet," as Cardano said of another ingenious contrivance, "unless it be the taking of Heaven by storm?"). *Bend* the open interval (0, 1) up into a semicircle and let the real line lie somewhere below it:

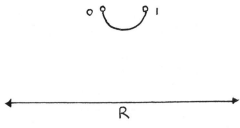

Now place your light source, or point of view, in that hollow bowl, midway between its missing end-points.

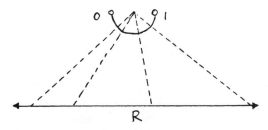

Each point of the open interval will correspond to a unique real, each real to a unique point of the interval. Since the interval and the real line are open (have no end-points), this match-up will work for every point on each. The cardinality of the reals is the same as that of any open interval of the reals.

The countably infinite was exemplified by many sets besides the eponymous counting numbers: the evens and odds, the multiples of any number, the integers and the rationals. Now this larger infinity is developing an entourage of its own: continuous open intervals of any size (analogous to those multiples of $\mathbb{N}$) as well as the reals altogether.

You need to pick out faces in the crowding natural numbers to jog your imagination into glimpsing just how very big their infinity is. Think of those largest known twin primes, with 29,603 digits in each; remember that there are primes thirty quintillion numbers apart with not another prime between them—yet there will be one further on. The number of naturals isn't just sickeningly huge: it is *infinite*. And yet, compared to the number of reals in the interval $(\frac{1}{3}, \frac{1}{2})$, it doesn't amount to a hill of beans. Worse: you've now seen that there are hills of reals in that narrow range compared to which all the hills of naturals look like valleys.

Does it take the courage of Daedalus or the foolishness of Icarus to ask now: "Is there a greater infinity still than those of the naturals and the reals?" Does the asking imply a sort of imagination in whose presence ours shrivels to a dot? Or has abstraction somehow insulated the mind against the reality it calls up, so that the imagination we rightly praise is one of intuiting relationships and devising ways of rigorously proving that they hold?

On January 5, 1874, Cantor wrote to Dedekind: "Could the points of a plane possibly be in a 1–1 correspondence with those on a line? Although the answer here too seems so obviously to be 'no' that you might almost think a proof superfluous, it seems to me that major difficulties stand in the way of an answer."

If he could only prove this, then a *third* size of infinity would be revealed—and then in all likelihood the points in space would constitute a yet larger, those in four dimensions a larger still, and so on forever, each higher dimension containing a greater infinity of points than the one before, as the spiral of counting widens and carries our thought out of the universe with it.

If he could only prove this . . . but no proof was forthcoming. What you often do in such a fix is to work simultaneously on proving *and* disproving your conjecture: one approach may suddenly prosper, or as each inches forward the odds against the other may suddenly lengthen. Picture, then, trying to set up a one-to-one correspondence between the points on the plane and those of the line—or to use the successful earlier tactic, between all the points in some neatly confined corner of the plane and part of the line—the "open unit square", perhaps, tucked into the first quadrant: all the points, that is, above the x-axis and below the line y = 1, and to the right of the y-axis and left of the line x = 1:

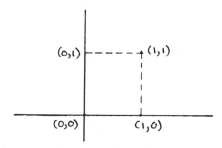

could these be unequivocally corresponded to the points in (0, 1) that this open square rests on?

The difficulty that seems insuperable is that each point of the square has two coordinates, and each point of the line only one. How could you find a unique point in (0, 1) to match up with a point in the square such as $(0.375\overline{0}, 0.9140286. . .)$? You clearly couldn't send this *pair* of reals just to its x-coordinate, for then every point in the square with the same x-coordinate would go there too and the correspondence would be far from one-to-one.

You couldn't send the two coordinates to their sum, since again many other points in the square would have coordinates that added up to the same value. Subtraction, multiplication, division of one coordinate by the other, or raising one to the power of the other all had the same fatal flaw.

Pressing the other line of attack seemed much more promising: assume a 1–1 correspondence and let it lead you to a contradiction. But this too went nowhere. Half a year later Cantor wrote to Dedekind again, asking him if he too was having difficulties with this—and adding that friends in Berlin said the whole business was absurd, since obviously two coordinates couldn't be matched up with one.

Three years went by. Cantor married, pushed his thoughts forward here and there, walked in the mountains—but this problem remained like a rock in a river, deflecting its currents without suffering any apparent erosion.

Then all at once, on June 20, 1877, he wrote to Dedekind that "despite believing the opposite for years," he had found a one-to-one correspondence between the points of a square and those of an interval. "*Je le vois mais je ne le crois pas*," he added nine days later: I see it but I don't believe it. Why did he declare this to his German friend in French? Was it ironic disarming, the way Beethoven called his most difficult and experimental efforts "Bagatelles"? Or was it meant to distance and elevate the revelation? Could it have been a tip of the hat to Descartes's staunch separation of seeing from believing? Or must we leave this as Churchill left Russia: a riddle wrapped in a mystery inside an enigma?

Once again you might reasonably expect that the correspondence Cantor found (or made) after so much time and with so much effort would be arcane almost to incomprehensibility, so that the marvel but not the meaning would reach us, as it would a medieval congregation listening to the Latin Mass. But his proof came from reaching backward rather than ahead: back to the idea of *interleaving* that let us count the integers, and back to a way of thinking so much younger than the sophistications of arithmetic as to be almost nonmathematical. He asks you to interleave the two coordinates of a point on the plane to find the one and only point of the line it will correspond to! So our example, $(0.375\bar{0}, 0.9140286\ldots)$, would go to the single decimal whose odd-numbered places are filled by the digits of $0.375\bar{0}$ and whose even-numbered places by those of the y-coordinate, $0.9140286\ldots$:

$$(0.375\bar{0},\ 0.9140286\ldots)$$
$$\downarrow$$
$$0.397154000208060\ldots$$

(Notice that we need the bar on the 0 of the x-coordinate in order to know that there will be only zeroes in all its decimal places from the fourth on.)

The correspondence Cantor made works in both directions, since any point in the interval (0, 1) will go to that point on the plane whose x-coordinate comes from the original point's odd-numbered decimal places, the y-coordinate from the even-numbered. So 0.29476583 corresponds to the point with coordinates (0.2468, 0.9753).

Dedekind wrote back immediately, congratulating his friend, but pointing out a technical problem which Cantor was later able to overcome (see the Appendix for the difficulty and its resolution). Diagnostic of the mathematician's faith in pattern was Cantor's postcard reply to Dedekind on June 23: "Unfortunately your objection is correct; fortunately it affects only the proof, not the conclusion." Aren't conclusions supposed to follow from proofs? Not if seeing has now replaced believing, making you *know* you are right. Proofs, like coats, can always be cut to fit your cloth. So in 1919, when Einstein received a telegram saying that astronomical observations had confirmed his theory of relativity, a doctoral student asked what he would have done had his predictions been refuted? "In that case," Einstein replied, "I'd have to feel sorry for God, because the theory is correct."

When Cantor published his revised proof, some—like the French mathematician Paul Du Bois-Reymond—objected that it was "repugnant to common sense." But Cantor had long since left the hearth of common sense to watch the aurora borealis of a distant sky, and brave were those both willing and able to follow him.

Since we know that $(0, 1) \leftrightarrow \mathbb{R}$, Cantor's proof meant that the cardinality of the square isn't greater than but the same as that of the reals. And larger open squares? As before, projection gives the 1–1 correspondence between all their points and those of the open unit square:

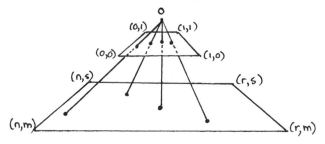

As for the open unit square and the whole two-dimensional plane, or $\mathbb{R}^2$, the cupping technique used in one dimension generalizes here:

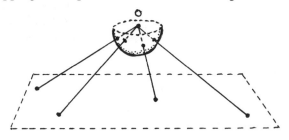

What then of all the points in three-dimensional space? Each such point has three coordinates (length and width away from the origin and height above it). Simply weave those *three* coordinates together in a triple braid:

$$(0.a_1a_2a_3a_4\ldots, 0.b_1b_2b_3b_4\ldots, 0.c_1c_2c_3c_4\ldots)$$
$$\downarrow$$
$$(0.a_1b_1c_1a_2b_2c_2a_3b_3c_3a_4b_4c_4\ldots).$$

So all of three-dimensional space has the same cardinality as the line. There are just as many points in the infinite universe as on the horizontal bar of this t.

And what about the imaginable but ungraspable points of four-, five-, ..., n-dimensional space: you can weave the four, five, or n coordinates of each of their points into a braid with the appropriate number of plaits, to get the same 1–1 match-up with the points on a line. "How many" has nothing to do with "how many dimensions."

This revelation startled Cantor as much as it does us. He had found two sizes of infinity, and as anyone who indulges in counting expects, where there are two there must be many more. Yet if the mind and the universe were divided, like Gaul, into three parts, he had conquered them all: the finite, the countable, and the continuous. You see why we said before that Cantor had done infinitely more than make sense of pairing numbers with their squares.

If you pause now to ask how he won his insights, the answer must surely involve a pioneer's love of freedom more than comfort. Three years of unrelenting work to prove the cardinality of the plane greater than that of the line—then giving it up on the spot because it wasn't true. He had freed himself from one more tenet of common sense, thanks to having already ruptured so many ligatures to traditional ways of thinking. Imagination is explosive—but if its explosions are to propel thought forward, they need to be confined, as are the explosions in an engine's cylinder. The best restraints aren't opinion but necessity: here, the need to replace belief by proof. "Mathematics is freedom!" Cantor later proclaimed—and freedom is where "what if" meets "what then".

You have probably come to terms with this about infinity: just when you think there's no more—there's more. In looking to higher dimensions for larger sizes, Cantor had turned in the wrong direction. What if greater cardinalities lay hidden back in the fundamental notion of set itself? He plunged into that mixture of doubting, defining, speculating, inventing, experimenting, redefining, and suddenly seeing one's work from afar, eternally there but newly discovered, which is the artist's inferno and paradiso. He emerged with the alephs.

For while a set is too primitive a notion to have any *structure*, with its elements just rattling around in it like marbles in a box, there is a shred of implicit architecture almost invisibly there too: the collection of its subsets. So a set S with three elements—call them a, b, and c—

$$S = \{a, b, c\}$$

brings with it the single-member subsets {a}, {b}, and {c}, and the three subsets with two elements each: {a, b}, {a, c}, {b, c}. Since a subset of S is any set T all of whose members are also in S, we should by courtesy include in this list Ø, the empty set (because, after all, whatever is in it is certainly also in S); and the set S itself, since it fits this definition. These last two subsets will remind you of the way definitions have stretched before, as in calling a single dot • a triangular, square, or any n-gonal number.

This set with *three* elements, then, has *eight* subsets:

$$Ø; \{a\}, \{b\}, \{c\}; \{a, b\}, \{a, c\}, \{b, c\}; \{a, b, c\}$$

$$1 \qquad 3 \qquad\qquad 3 \qquad\qquad 1$$

and eight is greater than three.

Is it always true that a set has more subsets than it has elements? Let's step back before leaping forward. If S = {a, b} it has *four* subsets:

$$Ø; \{a\}, \{b\}; \{a, b\}$$

$$1 \qquad 2 \quad 1$$

and if it has just *one* element, S = {a}, then it has *two* subsets:

$$Ø; \quad \{a\}$$

$$1 \qquad 1$$

Even the empty set, with *no* elements whatsoever, has one subset, namely, the set itself:

$$Ø$$

$$1$$

(people bothered by thinking of the whole set as a subset of itself sometimes distinguish this courtesy case by calling all the others "*proper subsets*").

If we put our results in order, what begins to take shape is the triangle Pascal invented to count combinations (though it was Jia Xian's triangle, six centuries before it was Pascal's—and al-Karaji's before Jia's . . .):

$$1$$
$$1 \quad 1$$
$$1 \quad 2 \quad 1$$
$$1 \quad 3 \quad 3 \quad 1$$
$$\cdot$$
$$\cdot$$

Here—if we call the top row *row zero*—the entries in the nth row add up to the number of subsets of a set with n elements.

A set with four elements, S = {a, b, c, d}, will have 1 empty set, 4 subsets with a single element in each, 6 with two members each, 4 with three, and the 1 "improper" subset with all the four elements:

$$1 \ 4 \ 6 \ 4 \ 1 = 16 \text{ subsets.}$$

You notice that if a set has n elements it will have 2^n subsets—and 2^n is always greater than n. You notice too that these are *all* the possible subsets of S. If you made up a peculiar rule for picking out some of its elements (such as choosing only those whose names begin with "v"), the subset they form will already be one of those we have listed by number of elements.

The pattern is beautiful in itself and spoke to Cantor in terms of cardinality: what he saw was that the cardinality of any finite set is less than that of the set of its subsets. This new set he called *the power set of S*, written $\mathscr{P} S$, in honor of 2^n, its cardinality.

If this were true of *infinite* sets, a cardinality greater than any he had so far discovered would spring out, and a greater after that, and a yet greater, forever: reviving the dream lost when spaces of higher dimension turned out to have the same cardinality as that of the real line. For the power set $\mathscr{P} \mathbb{R}$ of the reals—the set of all its subsets—would have more things in it than the uncountable number of reals; and then the set of all subsets of *this* new set $\mathscr{P}(\mathscr{P} \mathbb{R})$, the power set of the power set of the reals—would have more members still—world without end.

But why should the power set pattern continue for infinite sets—and even if it were to do so, how could we ever know it? In the face of such abstraction our capacities to prove seem (in the words of Daisy Ashford) as piffle before the wind. That Cantor in fact came up with a proof makes you wonder again about how impersonal the works of mathematics are. If the theorem is out there, is its proof out there too? Could anyone have discovered it, is it part of our common heritage? Or is the proof and what it establishes now a part of our thinking the way the Mona Lisa is—but needn't have been: necessary *after* the fact?

Once again Cantor approached his conjecture through a proof by contradiction—and once again he used his diagonal idea, but now etherealized to suit the unearthly remoteness of its subject. The spirit of this diagonal came to haunt all of his subsequent work—and after him became the controlling presence behind the foundations of modern mathematics.

We have just seen that the number of subsets outraces the number of elements in any *finite* set—but what if S were infinite? Well, assume that S in fact has exactly as many subsets as it has elements. That means you can make a 1–1 correspondence between them. Just as in the proof that the reals were uncountable, we can't specify beforehand what this correspondence is, since the proof must work for *any possible* correspondence. We are simply assuming that somehow each element of S matches up with one and only one of the subsets of S, and likewise that each of these subsets corresponds to a unique element of S. This will be true, we are assuming, whether S has as many elements in it as do the naturals, or as many as the reals—or in fact, has any cardinality whatever.

The empty set will appear in the list of subsets, of course, as will the whole set S, and the "singleton" subsets each with only one element, the subsets of all possible pairs, triples, and so on: they will all be there, along with the subsets formed in any way at all.

Trying to imagine such a correspondence is more mind-expanding than any drug in the hippie pharmacopoeia, but as a feeble help you might picture part of such a list as looking like this:

Elements of S		Elements of $\mathscr{P}$S: The Set of All Subsets of S
f	↔	$\{g, j\}$
g	↔	$\varnothing$
h	↔	S
i	↔	{every element in S except i}
j	↔	$\{g, h, j, l\}$
k	↔	$\{k\}$
l	↔	$\{g\}$
.	.	.
.	.	.

However the inventory is made, you will observe that an element of S either is or is not matched up with a subset of S that contains it (this is the same "diagonal" trope that lay behind the decimal made, in the earlier proof, of 5s and 6s). In our example, h, j, and k are matched with subsets containing them; f, g, i, and l aren't.

Now, says Cantor, all the elements of S that *aren't* in the subsets they correspond to on this putative list constitute a subset of S! They must: they form a collection drawn from the elements of S—and any such collection (no matter by what clear rule it is made) is a subset of S. Call this subset M. M must therefore appear somewhere on the list, matched up with some element of S: call it w.

Elements of S		Elements of $\mathcal{P}$S: The Set of All Subsets of S
.	.	.
.	.	.
w	$\leftrightarrow$	M
.	.	.
.	.	.

w must be a member of M or not; there is no third possibility.

But if w is an element of M, then it is in the subset it is matched up with. Yet M is the subset *only* of those elements in S that *aren't* in the subset to which they correspond. Hence w cannot be in M.

If it isn't in M, however, it isn't in the subset to which it corresponds on this list—and so it *must be* in M!

This contradiction shows that M (which seems like a perfectly good subset of S) isn't anywhere in the supposed 1–1 correspondence; yet it must be. *That* contradiction shows that such a correspondence is impossible, so the cardinality of $\mathcal{P}$S can't be the same as that of S. Since $\mathcal{P}$S contains at least as many subsets as there are elements in S (the singleton subsets in $\mathcal{P}$S are equinumerous with those elements), we can only conclude that the cardinality of $\mathcal{P}$S is greater than that of S (with 2^n in mind, we could write the cardinality of $\mathcal{P}$S as $2^{\text{cardinality of S}}$).

$$\infty$$

There are infinitely many counting numbers. There are yet more reals. Now there are more subsets of the reals than reals themselves. Take this new set $\mathcal{P}\mathbb{R}$: it begets a set with higher cardinality still—the set of all its subsets—and this iteration won't stop. Cantor had found new kinds of numbers and now needed to learn how they behaved. Each larger number is as perplexing as 7 is to the toddler who has just barely managed to distinguish 2 from 1. Is the next cardinality after that of a set S the size of its power set, $\mathcal{P}S$, and $\mathcal{P}(\mathcal{P}S)$ after that? Or are other infinite multitudes sprawled everywhere here, like titans grappling with primeval chaos?

Cantor's first task was to put his infinite cardinals in order, saying which was the least, which next—or to show that, like the fractions, none had a next larger after it. To do this he turned away from his cardinal numbers for more than a decade, working to find the calm mask of reason behind their unreadable faces. He still, as ever, acted with characteristic hubris.

Cardinal numbers tell us "how many"; *ordinal* numbers say where we are in an ordered sequence. So 1 is the smallest ordinal (think of it as "first") and the next comes by adding to it the unit "1", giving us 2 ("second"). The ordinals now go on sequentially by adding a unit to the latest one, making the third, fourth, and so on. It is a pleasant coincidence that each natural number plays two roles: as a cardinal it tells us how many; as an ordinal it tells us how far along the counting row we have come. This coincidence may disappear when we come to infinite sets.

Cantor looked at the endless sequence of finite ordinals and gave it a name: the last letter of the Greek alphabet, omega: ω. More than naming it, he thought of ω as *the first ordinal number after all of the finite ordinals!* Going in order he therefore counted: first, second, third . . . ωth— or as he wrote with forgivable ambiguity:

$$1, 2, 3, \ldots, \omega$$

where ω comes after *every* finite ordinal. This is somewhat beyond the wunnery tooery which came first for us. If you like, you may make his notion a little less uncomfortable by thinking of ω as standing for the natural order of the *whole set* of finite ordinal numbers. As such, it must come next.

Why not apply the generative principle of adding units once again, this time to ω, to get the *next* ordinal number, $\omega + 1$? This is reminiscent of the schoolyard exchange: "This is the gazillionth time I've won!" "Well, it's the gazillion and oneth time I've won!" "It's the infinitieth time I've won!" "Well, it's the infinity and oneth time I've won!" $\omega + 1$ is the ordinal number of a sequence with all the natural numbers in it (arranged, of course, in order), *and then* one element more—call it a fox or a cabbage or a goose, or simply "a"—which gets listed only after all the naturals:

$$1, 2, 3, \ldots, \text{fox}$$

or

$$1, 2, 3, \ldots, a$$

(this means that you take "2" out of the box after you've taken out "1", and "3" after "2" and so on—and take out the fox, or "a", only after you've taken out all the natural numbers. The mind reels). If instead we put in *two* extra elements, fox and goose or a_1 and a_2, after the natural numbers (with the understanding that $a_1 < a_2$), then $\omega + 2$ would be the ordinal number of the new sequence

$$1, 2, 3, \ldots, a_1, a_2.$$

We could go on like this—and Cantor did—getting in order the next ordinals $\omega + 3$, $\omega + 4$, and so on—until we come to a sequence made by inserting, *after* all the natural numbers, the elements $a_1, a_2, a_3, \ldots$—with as many subscripts as there are natural numbers. The ordinal number of *this* sequence

$$1, 2, 3, \ldots, a_1, a_2, a_3, \ldots$$

would be $\omega + \omega$.

You notice that ω and $\omega + \omega$ weren't themselves formed by adding a unit "1" to the ordinal immediately prior to each (since there wasn't one). What Cantor had done, really, was to bring in a second way of making ordinals alongside the normal one of adding a unit to the last so as to form the next. This second way took a great leap of the imagination—the very same leap that brought him to completed infinities. If the line of all the finite ordinals could be thought of as coiled in a box, then ω was the lid on this box. He spoke of ω in terms of *limit*, and from this point of view the new ordinals may now seem more comfortable still. ω is the limit which the finite ordinal numbers increase toward but never reach (just as $\frac{1}{2}, \frac{2}{3}, \frac{3}{4}, \frac{4}{5}, \ldots$ approaches but never reaches 1). And in fact, whenever there is no largest member in a succession of ordinal numbers, "then a new number is created," Cantor wrote, "which is thought of as the limit of those numbers, i. e., it is defined as the next number larger than all of them." Surely one of the most understated uses of "i. e." on record.

Now he could play his two generative principles off against one another to extend the ordinal numbers boundlessly beyond the finite into what he called the *Transfinite*.

$$1, 2, 3, \ldots \qquad \omega$$
$$\omega + 1, \omega + 2, \omega + 3, \ldots \qquad \omega + \omega$$
$$\omega + \omega + 1, \omega + \omega + 2, \omega + \omega + 3, \ldots \qquad \omega + \omega + \omega$$

$$\cdot$$
$$\cdot$$
$$\omega \cdot \omega = \omega^2$$
$$\omega^3$$
$$\cdot$$
$$\cdot$$
$$\omega^\omega$$
$$\cdot$$
$$\cdot$$
$$\cdot$$

But where are the *cardinal* numbers in all this splendor? We warned you that once past the finite, the pleasing coincidence of ordinal and cardinal might peter out—and it has. *Every one of these infinite sequences is countable*: it can be put, that is, into a 1–1 correspondence with the natural numbers! How can this be?

Take, for example, $\omega + 1$, which is the next number in order after ω—itself the first ordinal after the finite ordinals. $\omega + 1$ therefore stands, as you saw, for a set with *all* the natural numbers in it—and one element more. But counting "how many" is no respecter of order, so we may match up the elements in the sequence

$$1, 2, 3, \ldots, a$$

with the counting numbers by starting with a:

Ordinal	a	1	2	3 ...
	↕	↕	↕	↕
Cardinal	1	2	3	4 ...

The sequence whose ordinal number is $\omega + \omega$ we could count this way:

Ordinal	1	2	3	...	a_1	a_2	a_3 ...
	↕	↕	↕		↕	↕	↕
Cardinal	1	3	5	...	2	4	6 ...

$$\infty$$

The reason for Cantor's dozen-year digression into ordinal numbers was to bring order to the cardinals. What he discovered was that the *whole set* $\{\omega, \omega + 1, \ldots, \omega + \omega, \ldots, \omega + \omega + \omega, \ldots, \omega^2, \ldots, \omega^3, \ldots, \omega^\omega, \ldots\}$ was *not* countable—and it was the *first* uncountable set after each of its countable members (just as ω was the first *countably* infinite ordinal after each of the finite ordinals $1, 2, 3, \ldots$).

It was now that he reached for a name to distinguish these plateaus, these *sizes*, of ordinal—

$$\omega, \omega + 1, \omega + 2 \ldots$$

$$\vdots$$

$$4$$
$$3$$
$$2$$

and having used up the last Greek letter ω for them, he turned back for these transfinite cardinals to the first letter of the Hebrew alphabet, aleph: $\aleph$. As with everything he did, there were reasons behind the reasons for his choice. Aleph, as he said, itself represented "one" in Hebrew, and these new symbols marked a new beginning for his own work and for mathematics. Was there also here a private nod toward what he took to be the divine source of his inspiration? And was drawing from the language of the Old Testament, after borrowing from the Greek of the New, also a private acknowledgment of the Jewish background both of his converted ancestors and of his own Lutheranism?

The smallest transfinite cardinal—the size of such sets as $\mathbb{N}$ (or of a sequence like $1, 2, 3, \ldots, a_1, a_2$)—he called aleph null: $\aleph_0$. The next—for this ordinal substrate we have watched him rely on assured him that there *would* be a next—was $\aleph_1$: the size of the uncountable set $\{\omega, \omega + 1, \ldots, \omega^\omega \ldots\}$. The sequence of all these omegas Cantor denoted by an upper-case omega: Ω. He then considered the sequence

$$\Omega, \Omega + 1, \ldots, \Omega + \Omega, \ldots, \Omega^\Omega, \ldots$$

This is the first sequence to have more than $\aleph_1$ terms, so its cardinality is $\aleph_2$. Thus the Tower of Babel rises, calling forth on each new spiralled ledge names incomprehensible on the lower levels: $\aleph_3$, $\aleph_4$, and so on. Have you fully appreciated what "and so on" must mean? There will be an $\aleph_\omega$, and an $\aleph_{\omega+\omega}$, and an $\aleph_{\omega^\omega}$ ... How puny seem those infinitely large and larger fleas of Chapter Two:

> For at the gates that Cantor flung
> Apart (and Hilbert later)
> Angelic fleas cavort in hosts
> Inordinately greater.

Once again Cantor's faith in his ability to make sense of this new world had been justified. Whenever he had needed an insight or a proof, his inner voice had always accommodated. But supernatural helpers are notoriously unreliable: as the fisherman's wife in the fairy tale found out, ask once too often and the earth will tremble, the sky darken, and the sea run in mountainous black waves.

The first tremors that Cantor felt came from his colleagues. In France, Poincaré grew disgusted with set theory, which he thought pathological: he wrote that later generations would regard Cantor's work as a disease from which they had recovered. Closer to home, Kronecker, the most powerful figure in the German mathematical establishment, had been opposed to Cantor's ideas from the beginning, calling the work humbug and the man himself a charlatan and corrupter of youth. Kronecker tried repeatedly to prevent the publication of Cantor's papers, and his enmity kept Cantor in provincial isolation. An intellect focused on the infinite may overlook temporal indignities, but the psyche that intermediates between the intellect and the world cannot, and Cantor wrote that poverty and recrimination were the price he paid for his radical views. Mathematicians usually enjoy generalizing their observations, but seeing plots everywhere brought Cantor suffering. He alienated some friends, discarded others and had his first serious breakdown when he was thirty-nine.

It wasn't just scurrying intrigue or furtive academic cabals that the sky was darkening over, but the massive sliding and grinding of his thought's tectonic plates. When, a generation before, Schopenhauer had been dragged through long and embarrassing legal battles after throwing a seamstress down a flight of stairs, he consoled himself by reflecting that he was, after all, the author of *The World as Will and Idea*. But when Cantor looked at the world his will had created, he saw coastlines eroding.

There were fissures here and there that others would fill in time, such as proving that two sets have the same cardinality if and only if each is in 1–1 correspondence with a subset of the other. Cantor had to know such criteria if he were to arrange the alephs in order. And there was a problem he hardly noticed but which grew to monstrous prominence: even to show that $\aleph_0$ (the size of the set of counting numbers) was the smallest transfinite cardinal, an axiom was needed that sounds as innocent as the fisherman's appeal to the flounder for a pretty cottage by the sea instead of his hovel.

This *Axiom of Choice* asserts that given any collection of distinct, non-empty sets, if you need to (as Cantor did), you can always choose an element from each. It doesn't tell you how to do this, but you can imagine a jackdaw plucking a shiny trinket from each of a possibly infinite row of boxes.* The problem with the Axiom of Choice is that it lets the Four Horsemen loose in the land: it allows an initiate, for example (by an ingenious train of reasoning), to cut a golf ball into a finite number of pieces and put them together again to make a globe as big as the sun. Not only are its results an affront to intuition, but by not requiring us to know *how* we do the choosing, it adds to the Formalist shift of mathematics sideways. No longer do we construct objects cleated to their locales, but now rest content (like Hilbert and the Hungarian) with the bald assertion that they exist.

These commotions in the air were nothing compared to the mountainous clash of the waters that threatened to drown his world. For the whole point of propping up the cardinals on the ordinals was to find how the cardinals were arranged. Yet in doing so he might only have made a *different* sort of cardinal: those that corresponded to the ordinal plateaus. What had these to do with the cardinals that arose from the endless sequence of power sets? The problem is familiar: when Bombelli had come up with his "new kinds" of imaginary numbers, like $\sqrt[3]{2+\sqrt{-1}}$, he had no guarantee that these were of the same species as $a + bi$.

Stubborn, God-driven, isolated, heir of Alcibiades, Cantor insisted (proof came much later) that all transfinite cardinals *were* alephs—that is, they measured some stage in the growth of ordinals. He proved that the set of all subsets of N had the same cardinality as the set of the reals (the continuum)—for a proof, see the Appendix—and then claimed that this power set was the *next* aleph in order after $\aleph_0$. This was his famous *Continuum Hypothesis*: the cardinality of the continuum is $\aleph_1$.

Everything began to come apart. The Continuum Hypothesis obsessed him through recurrent breakdowns which took him in and out of sanitoria and his university's *Nervenklinik*. But the Continuum—for which a gothic C, as angular as his personality, has become the symbol—resisted his best efforts.

Some mathematicians thought there might be many alephs between $\aleph_0$ and the cardinality of C. Others attacked the whole ordinal enter-

---

*So to prove that there could be no transfinite cardinal less than $\aleph_0$, Cantor pictures any infinite subset A of N and asks us to go steadily through N until we come to the first element of A; then go through the subset of N made up of what's left until you come to A's second element—and so on. This 1–1 match-up with the counting numbers shows that A has cardinality $\aleph_0$ also.

prise: why must every infinite set have an aleph as its cardinal at all? And of those that did, how did we know that any two could be compared in terms of size? Jules König, renowned for his acuteness and reliability, announced at an International Congress in Heidelberg that he had proved the continuum had no aleph whatever corresponding to it. Cantor felt publicly humiliated, and although within a day König's proof was shown to rest on faulty assumptions, there was no still center now left to Cantor's turnings. When some colleagues met a few days later at Wengen to discuss the events of the congress, Cantor burst into the dining room of their hotel to explain excitedly to them—and everyone else at breakfast—just what had been wrong with König's proof.

$$\infty$$

It takes an inhuman force of character to make the beds while your house is falling down. At the same time that Cantor was trying desperately and unsuccessfully to prove the Continuum Hypothesis (if his theory couldn't even locate the cardinality of the reals in the hierarchy of the alephs, what good was it?), he went about the beginner's business of learning how to do arithmetic—but this time with transfinite cardinal numbers: the most radical extension of the franchise we have seen.

$\aleph_0 + \aleph_0 = \aleph_0$: the cardinality of the evens plus the cardinality of the odds is the same as the cardinality of the naturals. In fact—shades of Thabit ibn Qurra—any finite number of aleph nulls adds up to aleph null.

What was $\aleph_0^2$? Cantor knew the answer from the square array of the rationals, which he had shown was countable: $\aleph_0$ rows with $\aleph_0$ entries in each produced $\aleph_0$. Since you could likewise zigzag your way through a 3-dimensional array of rationals—or for that matter, an n-dimensional array—$\aleph_0^n = \aleph_0$. The size will increase when you move to the power set, $2^{\aleph_0}$; or to the next ordinal plateau—if those moves were different.

The same laws of addition and multiplication hold for any of the alephs: if k is any common ordinal, like 3 or 19—or even an infinite ordinal, like ω—$\aleph_k$ remains $\aleph_k$ when added to or multiplied by itself any number of times up to and including $\aleph_k$. Did Cantor's results come directly from his intuition, or from an abstract play of forms? "Is a man to follow rules—or rules to follow him?" asks Tristram Shandy, but assumes we know the answer. Cantor replied with a passage from Sir Francis Bacon: "We do not arbitrarily give laws to the intellect or to other things, but as faithful scribes we receive and copy them from the revealed voice of Nature."

This comes within an iota of the Intuitionist position that Brouwer was soon to establish. Was the iota that separated them no more than a different sense of "I"—Brouwer's the creator of mathematical reality, while Cantor (as he wrote to a friend in 1883) thought of himself as only a messenger, not the true discoverer of transfinite set theory? Yet no two mathematicians could be more unlike, since Cantor establishes existence on the basis of those proofs by contradiction that Brouwer abhorred. No wonder Hilbert so warmly defended Cantor's work: "No one shall drive us from the Paradise that Cantor has created for us!"

But was Cantor a Formalist? He never presented his results in the formal context of stripped-down deductions from axioms. His mathematics isn't about symbols that could mean this or that, but about what he saw as real ideas in the divine intellect, and corporeal objects in the world. Completed infinities were, for him, actual, not like the formless and merely potential *apeiron* of the Greeks. Hilbert purged mathematics of meaning; Cantor flooded his mathematics with metaphysics and theology.

Was this dissonance in his approach like trying both to prove and disprove a conjecture? Is his work the arena where revelation collides with language? Or was it once more a question of masks behind masks? For while the *form* of Formalism was absent from Cantor's writing, we see its spirit in his every line once we recall that this spirit is expressed by the Great Converse described in Chapter Two: what is consistent must exist.

"Mathematics," as Cantor had famously said, "is freedom!" But this motto is as ambiguous as it is bold, since there is freedom *from* as well as freedom *of*. For Cantor as for Hilbert, mathematics was free from contradiction: the coherence of its parts in a consistent whole was all the proof you needed that the whole existed. Should intuition give out (as it does when we think about the higher alephs), proofs *by* contradiction would take us to results purified *of* contradictions.

$$\infty$$

A foolish consistency may be the hobgoblin of little minds, but consistency itself of great ones. What if irreparable paradoxes were now to open up in the fabric of Cantor's work? Then the mountains would fall and the sea roll over the land.

The melodiously named mathematician Cesare Burali-Forti, at the Military Academy in Turin, discovered in 1897 a curious consequence of the new set theory. Take the sequence of *all* the ordinal numbers. Since this sequence is itself ordered, it must have an ordinal number—let's call it J. J would have to be greater than any of the ordinals in the sequence (it is their successor or limit)—yet these are *all* the ordinals, so J must both be and not be among them. There this paradox squats, as complete and immovable as the sphinx with its riddle. You can't get under, over, or around it.

The repercussions of Burali-Forti's paradox were immediate, far-reaching, and devastating. Nothing had been more certain than mathematics; now, said a contemporary, nothing had become so uncertain. The only certainty, surely, was that this blow would definitively topple Cantor's precarious mental balance. Far from it. He announced that he had discovered this paradox two years before, and thought of it as a positively beneficial result. In fact he added another to it: the set of all sets—call it S—would, being a set, have a certain cardinality. But the set of all its subsets, $\mathcal{P}S$, would have a greater cardinality still. Since S is the set of *all* sets, $\mathcal{P}S$ must be an element of it, so that a set of higher cardinality would be contained in a set of lower cardinality.

You may think Cantor's welcoming of the paradoxes was either bravado in the face of defeat,

> Thought shall be harder, heart the keener,
> Courage the greater, as our might lessens

(as the Anglo-Saxon writer of the *Battle of Maldon* put it), or a sign indeed of breakdown. But there were always masks behind his masks. It is telling that on the terrible day of public shame, when it seemed that König had destroyed any hope for the Continuum Hypothesis, Cantor thanked God for reproving him for his errors *and at the same time* asserted that König's demonstration would be found erroneous.

Cornered by the paradoxes, Cantor's thought twisted and turned them seemingly to his advantage. He made out a new distinction between *consistent* and *inconsistent* collections: only the former were sets; the latter (such as the collection of all sets or of all the ordinals) were not. Whatever they were, their existence at last let him prove (he wrote to Dedekind) what he vitally needed to know: that sets could have no cardinality other than the alephs. This meant that his two different ways of making larger and larger sets—via power sets or via the cliffs, no-man-fathomed, among the ordinals—coincided in what mattered: the way their size was measured.

You will find the proof he sent to Dedekind in the Appendix. Consider here, instead, the implications of his decision to use the very inconsistency of collections that were "too large" in order to establish facts about consistent collections—that is, about sets. These inconsistent collections Cantor calls the *absolutely infinite*: it was the infinite that God alone could know, as if our seeing it as inconsistent was a reproach to our feeble humanity for daring to extend its thought so far.*

---

*Doesn't "the heaven of heavens" re-echo in the set of all sets? "But will God in very deed dwell with men on the earth? Behold, the heaven of heavens cannot contain Thee; how much less this house which I have built!" (2 Chron. 6:18). Awe, and its obverse, humility, give religion its endlessly ordinal impulse.

The hierarchy of larger and larger infinities that Cantor had reported on he called only *Transfinite*, precisely to distinguish such things that humans could think of (approaches to the Throne) from what lay exclusively in the mind of God.

Yet at the same time that Cantor's piety and humility showed themselves in this distinction, he pushed his own thought past that brink to make the absolutely infinite reveal new truths about the transfinite: to guarantee the validity of his discoveries.

His leap upward from towering heights to bring back knowledge of the lower structure has characterized subsequent work in set theory, a century now and more after that moment. An active research program strives to gain insight into very finite situations by invoking transfinite numbers that dwarf even Cantor's remoter alephs. There is the First Inaccessible Cardinal and after it, Hyper-Inaccessibles; then the First Mahlo Cardinal (sounding more like a medieval grandee than a size past the Hyper-Inaccessibles themselves). There are Cardinals Indescribable, Huge, Supercompact; Rowbottom and (it may be) Ramsey Cardinals, and then the Extendible and perhaps the Ineffable Cardinals, not to mention those that are Inexpressible. Devising them isn't only a game of one-upmanship on a gigantic scale, but a serious attempt to prove important theorems which are *unprovable* without their condescending help: fetching from afar carried to its logical extreme.

In the fairy tale the sky had turned as black as pitch and the fisherman had to shriek out to make the flounder hear that his wife now wanted to be Lord of the Universe.

> "Now she must go back to her old hovel," said the flounder, "and there she is!" So there they are to this very day.

In what passes for the real world, no one could make sense of Cantor's proof. Long after, Zermelo said of it that "the intuition of time is applied here to a process that goes beyond all intuition, and a fictitious entity is posited of which it is assumed that it could make *successive* arbitrary choices"—over a span longer than time's.

And now it wasn't just that certain collections were simply too large to be consistent: in 1907 Bertrand Russell showed that set theory generated paradoxes that hadn't anything to do with alephs or ordinals at all. He had found that legitimate ways of defining a set (via the properties shared by its members) led to nonsense. Take, for example, the set of all those sets which aren't elements of themselves (an example of one of those sets would be the set of all apples, which isn't itself an apple). Is the set of all such sets an element of itself? If it is, then (by its very definition) it isn't; but if it isn't, then (again by the way it is defined) it is.

Sets were so intuitively clear when we began that we were happy to reduce the mysteries of number to them. Now they have uncontrollably swollen and multiplied, and which are consistent, which inconsistent, and for whom? Their very nature has grown incoherent.

Cantor asked Dedekind how he thought of a set. Imagine them walking together in their beloved Harz mountains, where new and wonderful vistas opened at every bend. Dedekind said that for him a set was a closed bag with specific things in it which you couldn't see and knew nothing about, except that they were distinct and really there. A few minutes passed. Cantor, immensely tall, flung out his arm toward the wild landscape: "A set," he said, "I think of as an abyss."

∞

The question of infinity had brought mathematics to the edge of uncertainty.

—Joseph Warren Dauben

Cantor struggled doggedly to prove his Continuum Hypothesis, that the cardinality of the continuum was $\aleph_1$. Elation alternated with longer and longer depressions lit fitfully by promising strategies that one after another flickered out. He was hospitalized again and again. Why had he ever as a young man given up music for mathematics, he now wondered, recalling the days when he had played the violin and formed his own string quartet. Having broken with so many of his colleagues over the years, he continued to thank his wife for each dinner she provided and to ask her at its end whether she still loved him.

He began to concern himself with the Rosicrucians, and Theosophy, and Freemasonry—and with proving that Shakespeare's plays had really been written by Sir Francis Bacon. He hinted darkly that he had made certain discoveries concerning the first king of England "which will not fail to terrify the English government as soon as the matter is published."

Cantor, a few months before his death.

Form always seeks substance, and in doing so begets ever more shadowy forms. Sets behind numbers, inconsistent collections behind sets, ordinals behind cardinals, the absolutely infinite behind the transfinite, his father's voice

261

in the background and a secret, divine voice behind that. . . . Cantor published a pamphlet, "Ex Oriente Lux", revealing that Christ was the natural son of Joseph of Arimathea, and in this confusion of mask after mask died, aged seventy-three, in 1918. His great work on set theory of 1883 was prefaced with three quotations, the last of which was, "The time will come when these things which are now hidden from you will be brought into the light."

What has since been revealed about Cantor's Continuum Hypothesis is of a piece with this endless and endlessly surprising drama, shaped by spiral returns to the earliest days of our story. Then, you recall, Hippasus took the very means created by his teacher, Pythagoras, to undermine the Pythagorean cosmos. His readiness to see askew what others had looked at head-on led him to find in the diagonal of a square the irrationals that now swarmed among the ratios which alone were supposed to resound in the physical, mental, and moral order of things. It was precisely Cantor's daring diagonal which Kurt Gödel turned round to prove that there were more true propositions than proofs: that in any sufficiently rich formal system there would be statements which could neither be proved nor refuted. Gödel's own later work, and that of the American logician Paul Cohen, then showed that the Continuum Hypothesis was one of these statements. It wasn't just that Cantor couldn't prove it; this time the difficulty did lie in the problem itself: no proof or disproof lodged anywhere within the arcaded city of formal set theory. It was forever undecidable.

Mathematics is permanent revolution. Gödel's inevitably followed from the radical mathematics invented by Cantor.

> Revolutions still more remote appeared in the distance of this extraordinary perspective. The mind seemed to grow giddy by looking so far into the abyss of time . . . and we became sensible how much further reason may sometimes go than the imagination can dare to follow.

John Playfair wrote this at the beginning of the nineteenth century about the Great Unconformity at Siccar Point, which the geologist James Hutton had all at once seen as slanted layers of time, thus changing forever our view of the earth's evolution and ultimately ours. Cantor's transfinite arithmetic is this Unconformity on a universal scale. It has disrupted our sedate understanding of the mind and its world, and from its fracture a new understanding has yet fully to emerge. When it does—when the doors of our perception are finally cleansed, as William Blake promised—then everything will appear as it is: infinite.

But which infinity will we see?

Appendix

1. [to page 45] An inductive proof that the sum of the first n odd integers is n^2.

To prove that $1 + 3 + 5 + \ldots + (2n - 1) = n^2$

1. <u>Prove that the statement is true for n = 1.</u>
   If n = 1, the only odd integer to consider is the first: 1. And is it true that when n = 1, n^2 = 1? Yes.
2. <u>Assume the statement is true for n = k.</u>
   Easily done. Since when n = k the kth odd number is 2k – 1, we will be assuming that

$$1 + 3 + 5 + \ldots + (2k - 1) = k^2.$$

3. <u>Now, using this assumption, prove that the statement is true for k's successor, k + 1.</u>

Since the $(k + 1)^{th}$ odd number is 2 more than the kth odd number, we know that it is 2k + 1. So we want to prove that

$$1 + 3 + 5 + \ldots + (2k - 1) + (2k + 1) = (k + 1)^2.$$

But we know from our assumption in step 2 that everything on the left-hand side up through (2k – 1) is equal to k^2, so that all we have left to prove is that

$$k^2 + (2k + 1) = (k + 1)^2.$$

Squaring the right-hand side gives us $k^2 + 2k + 1$, so whipping the parentheses off the left-hand side reveals that we do indeed have our equality.

2. [to page 50] Models and consistency.

An accurate model will have the same *structure* as what it represents, even if its size, appearance, and material are entirely different. So Hilbert's last broadcast words and his laugh at the end of them are faithfully modelled by the bumps and dips of the grooves in the recording made at the time.

Hilbert made inspired use of this simple idea. He borrowed or built *within* Euclidean geometry an accurate model of each of its rivals. Hence, if Euclidean geometry turned out to have no contradictions in it, neither could they. And then he showed how—as with Descartes's coordinate geometry—to make a model of Euclidean geometry within arithmetic. Everything therefore now hinged on showing arithmetic to be consistent.

You may wonder how a model could be made within Euclid's of a geometry that violated his parallel postulate—one in which there were many parallels, for example, to a given line through a point not on it, rather than one. The cleverness lay in thinking of familiar objects in unfamiliar ways. Take the interior of a circle as this non-Euclidean geometry's whole two-dimensional universe, and some chord in it as the "given line". Through some point inside the circle but not on this line there will be many chords that never *within the circle* intersect the given one, and are therefore parallel to it—in this model.

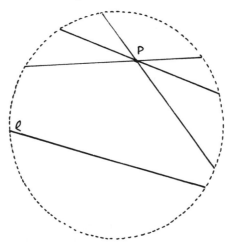

As a step toward proving that the Peano axioms for arithmetic were consistent, Hilbert proved that the field axioms on page 38 were. This again he did by making an elegant little model of them. Since this microcosm existed—there it was, right under the mathematical eye—and nothing that exists can harbor a contradiction (nothing can both be and not be in something that is), the field axioms had to be consistent.

This was his model. The only numbers it contained were 0 and 1, and addition and multiplication were *defined* by the following tables (since "addition" and "multiplication" are primitive terms, we can model them as we will—so long as they still obey the axioms):

+	0	1		x	0	1
0	0	1		0	0	0
1	1	0		1	0	1

If you check you will see that this two-element universe satisfies all of the field axioms.

Such a universe cannot, however, satisfy the more demanding Peano Axioms (look at axiom 5). Hilbert's vain attempts to model those turned out not to be a personal but a formal failing, which was what Gödel showed in 1931 (a logician named Post was in fact prior, but hadn't published his results): induction made it impossible to prove *within* such a system that it was both consistent and complete—and that is pretty much how things still stand.

It was ingenious to think of making models for a world that is generated by a set of axioms, in order to show that these models were machines that would go of themselves. It may lead you, however, to a curious speculation. Aren't the axioms supposed to catch the essence of whatever piece of mathematics they describe? If the real numbers, for example, are truly real—part (or all!) of the one and only universe—how could they have many models which are all but structurally different: different in material, appearance, and even in their number of parts? The one shown above comes from a crowd more numerous and motley than any that surges through one's minimalist imaginings. It is the Schumann problem come back to haunt us: you feel the utter uniqueness of your true, your subjective self (apprehended perhaps by intuition), distinct from the body it happens to inhabit—a body whose form (even to fine details) you share with billions of different members of our species. If mathematics is formal through and through, then it will be no surprise (only a disappointment) that it doesn't pick out or point to what *really* exists, in the sense of being uniquely distinct in time.

In an ancient tale from India, the gods—each eager to win the hand of beautiful Damayanti—turn themselves into perfect simulacra of Nala, her human beloved, and stand in an endless row with him in their midst. She may marry him only if she can single him out from the models. This she does in an instant, since his feet alone touch the earth. No formal incarnation of an axiom system seems to be similarly grounded. True to its Romantic origins, Formalism lets loose doubles and doppelgängers, mimics and masks, in a fun-house of distorted reflections.

If rather than being formal itself, mathematics is *about* form—about the unique patterns into which relations must fall—then any approach to it that blurs this uniqueness must be at best a sort of scaffolding trundled up by one of the blind sages against some part of the elephant— the combinatorial play of too few assertions launched against the world's endless subtlety.

That laugh of Hilbert's happened at an instant of time. Models of it may soon make their way to you, via http://topo.math.u-psud.fr/~lcs/ Hilbert/HlbrtKD.htm.

To Chapter Three

1. [to page 56] Proofs of some fundamental propositions.

We'll prove that 0 is the only additive identity. The proofs of the other statements have the same form.

Assume, as now seems the natural way to begin, that there *is* another additive identity—call it ö. Then ö + a = a and 0 + a = a also. Hence ö + a = 0 + a.

Now add –a to both sides of the equation:

$$(\ddot{o} + a) + {-a} = (0 + a) + {-a} .$$

Group together the second and third terms (by Associativity):

$$\ddot{o} + (a + {-a}) = 0 + (a + {-a})$$

and a + –a = 0 by the Additive Inverse Axiom, giving us

$$\ddot{o} + 0 = 0 + 0 .$$

By the Additive Identity Axiom, this becomes

$$\ddot{o} = 0 ,$$

so ö was just 0 wearing a mask.

2. [to page 58] A more rigorous proof that a negative times a positive is negative.

Put $(-a) \cdot b$ in a helpful context:

$$(-a) \cdot b + a \cdot b = b \cdot (-a + a)$$

by the Commutative and then the Distributive Axiom. But $b \cdot (-a + a)$ is $b \cdot 0$, which we now know is 0. So

$$(-a) \cdot b + a \cdot b = 0 .$$

This means that $(-a) \cdot b$ is the additive inverse of $a \cdot b$. But the additive inverse of $a \cdot b$ is $-(a \cdot b)$: hence,

$$(-a) \cdot b = -(a \cdot b) ,$$

since each number has a unique inverse. This brief whirl around the floor brings us to the conclusion that a negative times a positive is negative.

3. [to page 59] A visual proof that a negative times a negative is positive.

The following visual proof that $(-a)(-b) = ab$ relies on the properties of similar triangles we spoke of on page 15, when we showed how to think of multiplication visually on the Euclidean plane. Here, however, the picture lies on the somewhat more artificial coordinate plane, where negative numbers are represented by horizontal lengths in the second and third quadrants and vertical lengths in the third and fourth.

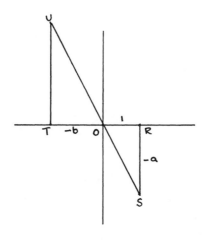

Appendix

Lay off a line of length 1 on the x-axis, from $(0, 0)$ to $(1, 0)$, and call the end-points O and R. Now draw a perpendicular from R down into the fourth quadrant, ending at S, representing the number –a; and on the x-axis, from O leftward to T, lay off a segment representing –b.

Draw a line through S and O upward into the second quadrant, meeting the perpendicular from T at U. Since $\triangle ORS \sim \triangle OTU$, $\frac{TU}{-b} = \frac{-a}{1}$, so that $TU = (-a)(-b)$, but is positive, since it is a vertical in the second quadrant.

4. [to page 67] Factors of terms in the sequence 2, 5, 8, 11, 14, . . .

To ask about which primes are in the sequence

$$2, 5, 8, 11, 14, 17, \ldots$$

means asking about what factors each term in it could have. The choices can only have the form $3n - 1$, $3n$, or $3n + 1$. The factors of any term here couldn't all have the form $3n$, because if you multiply such numbers together you would get another of the form $3n$, not $3n - 1$:

$$3a \cdot 3b \cdot 3c = (27 \cdot abc) = 3 \cdot (9abc).$$

Nor could the factors all be of the form $3n + 1$, because

$$
\begin{array}{r}
3a + 1 \\
\times\, 3b + 1 \\
\hline
9ab \;+ 3b + 3a + 1 \\
= 3(3ab +\ b\ + a) + 1,
\end{array}
$$

which is another number of the form $3n + 1$. Were some factors of the form $3n$ and the rest $3n + 1$, their product would again be of the form $3n$. We have found, therefore, that since the numbers in our sequence have the form $3n - 1$, at least one of the prime factors of any number in our sequence must have the form $3n - 1$.

5. [to page 70] e and its logarithm.

Some functions may never stop growing, but grow at different rates from one another. Among the rapid risers are the *exponential functions*, where the variable is used as an exponent, like $f(x) = 10^x$: as x increases, the output of 10^x rockets away. Here are 2^x and 3^x; try graphing 10^x yourself, and be daring about which values for x you use as inputs: your calculator will let x take on any real value, not just integers.

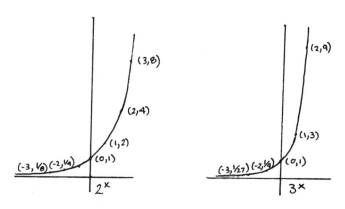

As you can see, 3^x rises more steeply than 2^x: it grows more rapidly. Between 2 and 3 lies a remarkable number, named e by Euler in 1728 (e is irrational, and is approximately 2.718281828459045...), which has the important property that e^x grows at precisely the rate, for any x, of the function's *output* at that x. While with all such exponential functions the more you have, the more you get; with e^x how much you have is *exactly* how fast you grow. Since this so well describes organic growth, $f(x) = e^x$ shows up everywhere in the biological world, from the growth of cells to the growth of animal populations. Its graph looks like this:

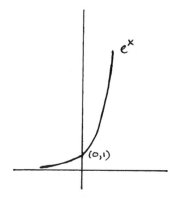

The output of $f(x) = e^x$ tells you how big e^x is for a given input x. What if you wanted to answer, however, the paired question: what x must I put in to get a certain output from this function? This is answered by its paired function, $g(x) = \ln x$ (which stands for "the logarithm, with base e, of x": ponderous name for a svelte idea).

Since you need to raise e to about 2.30258... to get 10 (so $f(2.30258) = e^{2.30258} \approx 10$, where "$\approx$" means "is approximately"), we would say:

$$\ln 10 \approx 2.30258$$

The graph of this "natural logarithm" function is the mirror-image of the graph of f(x) = e^x,

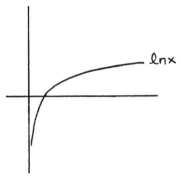

where the mirror is the line slanted at 45° between them: the line y = x (because we're exchanging the roles of input and output; and since this line is where they are equal, the two graphs will be symmetrical around it):

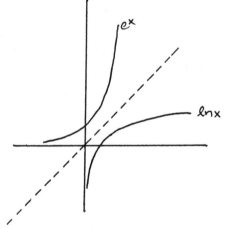

Not the least surprising thing about the relevance of e^x and ln x to the world of primes is this importing of the organic into what seemed mechanical. Are the two more intercalated than we supposed, the distinction between them as artificial as that between the animal and vegetable kingdoms has turned out to be?

To Chapter Four

1. [to page 98] The series of reciprocals of primes.

Euler's proof that this series in fact diverges makes use of sophisticated techniques. We will follow instead a delightful proof from 1966 by James Clarkson.

The effort it will take here and there, and the overall pull, can only strengthen your muscles for mathematics. Since the chief difficulty will be keeping the gist of the proof in focus at times when the point of particular moves isn't clear, patience is at a premium. It will also help to step back from it now and then and review in your mind what has already happened and what the grand design is. As Brouwer knew, most mathematics is done with the eyes closed—or as people in math libraries can tell you, when someone is staring at a book, he isn't doing mathematics; when he is staring at the ceiling, he is.

You know from Chapter Two that every natural number n has a multiplicative inverse, $\frac{1}{n}$, also called its reciprocal. The series we want to consider is the sum of the reciprocals of all the primes:

$$\frac{1}{2} + \frac{1}{3} + \frac{1}{5} + \frac{1}{7} + \frac{1}{11} + \frac{1}{13} + \frac{1}{17} + \ldots$$

Since 2 is the first prime, let's call it p_1. 3 is the second prime, p_2; $p_3 = 5$, $p_4 = 7$, and so on. The subscripts correspond to the order the primes come up in. We can therefore write:

$$\frac{1}{p_1} + \frac{1}{p_2} + \frac{1}{p_3} + \frac{1}{p_4} + \ldots$$

It will be handy to abbreviate this infinite sum by a concise symbol, since we will have much to do with it. The pleasant convention is to use the upper-case Greek sigma, Σ, for "sum", followed by what it is a sum of—in this case—$\frac{1}{p_i}$, where that index i runs from 1 through all the natural numbers:

$$\Sigma \frac{1}{p_i}.$$

To show that i begins at 1 and goes on to infinity, we decorate Σ with i = 1 below and ∞ above, so that

$$\sum_{i=1}^{\infty} \frac{1}{p_i} \text{ means } \frac{1}{p_1} + \frac{1}{p_2} + \frac{1}{p_3} + \ldots \left(\text{i.e., } \frac{1}{2} + \frac{1}{3} + \frac{1}{5} + \frac{1}{7} + \ldots \right).$$

Clarkson's strategy is to assume that this sum converges and then to get a contradiction: namely, that another series which we know *diverges* would then have to converge too. That other series is based on the harmonic series:

$$\frac{1}{1} + \frac{1}{2} + \frac{1}{3} + \frac{1}{4} + \ldots$$

(which we can now abbreviate $\sum_{n=1}^{\infty} \frac{1}{n}$). We saw on page 95 that this series diverges.

A clever tactic Clarkson uses depends on noticing that a divergent series multiplied by a positive fraction still diverges: since $\sum_{n=1}^{\infty} \frac{1}{n}$ grows greater than any particular number, so will a third or a fourth of it, or $\frac{1}{A}$ times it (where A is some positive constant), that is,

$$\frac{1}{A} \sum_{n=1}^{\infty} \frac{1}{n} \text{ diverges too.}$$

Since "$\frac{1}{A} \sum_{n=1}^{\infty} \frac{1}{n}$" means $\frac{1}{A}$ multiplied by each of the terms in the series, we can write it as:

$$\frac{1}{A} \sum_{n=1}^{\infty} \frac{1}{n} = \frac{1}{A} + \frac{1}{2A} + \frac{1}{3A} + \ldots$$

The trick is to find the right A which will tie the reciprocals of primes to the terms of the harmonic series. Here he makes use of a very nice idea. If you have two fractions, like $\frac{1}{5}$ and $\frac{1}{11}$, the new fraction $\frac{1}{5 \cdot 11} = (\frac{1}{5})(\frac{1}{11})$ will be less than $(\frac{1}{5} + \frac{1}{11})^2$, because

$$\left(\frac{1}{5} + \frac{1}{11}\right)^2 = \left(\frac{1}{5}\right)^2 + 2 \cdot \left(\frac{1}{5}\right)\left(\frac{1}{11}\right) + \left(\frac{1}{11}\right)^2 .$$

In the same way, given three fractions, say $\frac{1}{P_8}$, $\frac{1}{P_{13}}$, and $\frac{1}{P_{19}}$, $\frac{1}{P_8 \cdot P_{13} \cdot P_{19}} < (\frac{1}{P_8} + \frac{1}{P_{13}} + \frac{1}{P_{19}})^3$, since $\frac{1}{P_8 \cdot P_{13} \cdot P_{19}}$ will be among the terms that arise from cubing the sum of those three fractions.

In fact, and for the same reason, if you have r different fractions $\frac{1}{n_1}$, $\frac{1}{n_2}$, $\frac{1}{n_3}$, ..., $\frac{1}{n_r}$, then

$$\frac{1}{n_1 \cdot n_2 \cdot n_3 \cdot \ldots \cdot n_r} < \left(\frac{1}{n_1} + \frac{1}{n_2} + \frac{1}{n_3} + \ldots + \frac{1}{n_r}\right)^r .$$

$\frac{1}{n_1 \cdot n_2 \cdot n_3 \cdot \ldots \cdot n_r}$ will be even less, of course, than that sum plus a great many more fractions, raised to the power r. So for example $\frac{1}{P_8 \cdot P_{13} \cdot P_{19}}$ will be very much less than the sum of *all* the reciprocals of primes from the eighth prime on, cubed:

$$\frac{1}{P_8 \cdot P_{13} \cdot P_{19}} < \left(\frac{1}{P_8} + \frac{1}{P_9} + \frac{1}{P_{10}} + \ldots + \frac{1}{P_{13}} + \ldots + \frac{1}{P_{19}} + \ldots\right)^3 .$$

This idea is one of the two sticks Clarkson will rub together to spark into existence that A he needs.

The other stick is this. If $\sum\limits_{i=1}^{\infty} \frac{1}{P_i}$ converges (as we are assuming), then its limit—its total sum—will be a certain number L. As the terms of the series add up, their sum will get closer and closer to L—and as it does, it will grow to within 1 of that limit, and then to within $\frac{1}{2}$ of it. In other words (and this is what he is after), there will be some prime—the kth along the way, p_k—such that

$$\sum_{i=1}^{k} \frac{1}{P_i} = \frac{1}{P_1} + \frac{1}{P_2} + \frac{1}{P_3} + \ldots + \frac{1}{P_k}$$

will be less than $\frac{1}{2}$ away from L:

$$\frac{1}{P_1} + \frac{1}{P_2} + \frac{1}{P_3} + \ldots + \frac{1}{P_k}$$

$$L - \frac{1}{2} \qquad L$$

This means that the sum of *all the rest* of the terms, from the next one, $\frac{1}{P_{k+1}}$ on, will add up to less than $\frac{1}{2}$:

$$\sum_{i=k+1}^{\infty} \frac{1}{P_i} = \frac{1}{P_{k+1}} + \frac{1}{P_{k+2}} + \ldots < \frac{1}{2}.$$

Why choose $\frac{1}{2}$? Why does $\frac{1}{2}$ matter? Because Clarkson has in the back of his mind something else that we discovered in Chapter Four:

$$\sum_{t=1}^{\infty} \left(\frac{1}{2}\right)^t = \frac{1}{2} + \frac{1}{4} + \frac{1}{8} + \ldots$$

converges (in fact, it converges to 1: see page 90).

Now we can start to rub his two sticks together. The first task is to find that A we need.

Assuming that $\sum\limits_{i=1}^{\infty} \frac{1}{P_i}$ converges meant that there would be a k such that $\sum\limits_{i=k+1}^{\infty} \frac{1}{P_i} < \frac{1}{2}$. Well, take those first k primes and multiply them together—let's call their product Q:

$$P_1 \cdot P_2 \cdot \ldots \cdot P_k = Q.$$

Since every one of these k primes is a factor of Q, it is a factor as well of nQ, where n is any positive integer. This means that none of the first k primes can be a factor of 1 + nQ (for if it were, it would have to be a factor of 1 as well, which is impossible—an echo of Euclid's proof that there is no last prime).

So the factors of $1 + nQ$ must lie among the primes beyond p_k: among p_{k+1}, p_{k+2}, and so on. In other words, $1 + nQ$, for each integer n, is a product of primes of the form p_{k+m} (where m is an integer ≥ 1).

If, for a particular n, $1 + nQ$ is a product of s different primes of this form,

$$1 + nQ = \underbrace{P_{k+a} \cdot P_{k+b} \cdot \cdots \cdot P_{k+s}}_{s \text{ of these}}$$

then

$$\frac{1}{(1+nQ)} = \frac{1}{(P_{k+a} \cdot P_{k+b} \cdot \cdots \cdot P_{k+s})}.$$

Does this look familiar? Yes; we saw something very much like it on page 272, where we found (in terms of r different numbers $n_1, n_2, \ldots, n_r$) what we could here express in terms of s different primes $P_{k+a}, P_{k+b}, \cdots, P_{k+s}$:

$$\frac{1}{P_{k+a} \cdot P_{k+b} \cdot \cdots \cdot P_{k+s}} < \left(\frac{1}{P_{k+a}} + \frac{1}{P_{k+b}} + \ldots + \frac{1}{P_{k+s}} \right)^s.$$

As we said before, this will be even less than the sum of *all* the reciprocals of primes from $\frac{1}{P_{k+1}}$ on, that sum raised to the power s:

$$\frac{1}{(1+nQ)} = \frac{1}{(P_{k+a} \cdot P_{k+b} \cdot \cdots \cdot P_{k+s})} < \left(\frac{1}{P_{k+a}} + \frac{1}{P_{k+b}} + \ldots + \frac{1}{P_{k+s}} \right)^s <$$

$$\left(\frac{1}{P_{k+1}} + \frac{1}{P_{k+2}} + \ldots \right)^s.$$

This gives us the A we want: let $A = 1 + Q$. For then watch what happens (keep in mind that if $x > y$ then $\frac{1}{x} < \frac{1}{y}$): if $\frac{1}{A} = \frac{1}{(1+Q)}$, then

$$\frac{1}{A} \sum_{n=1}^{\infty} \frac{1}{n} = \frac{1}{A} + \frac{1}{2A} + \frac{1}{3A} + \ldots$$

$$= \frac{1}{(1+Q)} + \frac{1}{(2+2Q)} + \frac{1}{(3+3Q)} + \ldots$$

274

But $2 + 2Q > 1 + 2Q$ and $3 + 3Q > 1 + 3Q$—in fact,

$$n + nQ \geq 1 + nQ \quad \text{so} \quad \frac{1}{(n+nQ)} \leq \frac{1}{(1+nQ)},$$

$$\text{i.e.,} \quad \frac{1}{(1+nQ)} \geq \frac{1}{(n+nQ)}$$

(with equality only when n = 1), so

$$\frac{1}{(1+Q)} = \frac{1}{(1+Q)}$$

$$\frac{1}{(1+2Q)} > \frac{1}{(2+2Q)}$$

$$\frac{1}{(1+3Q)} > \frac{1}{(3+3Q)}$$

$$\cdot$$
$$\cdot$$
$$\cdot$$

Adding them all up,

$$\sum_{n=1}^{\infty} \frac{1}{1+nQ} > \sum_{n=1}^{\infty} \frac{1}{n+nQ} = \left(\frac{1}{1+Q}\right) \underset{\uparrow}{\sum_{n=1}^{\infty} \frac{1}{n}}.$$

the harmonic series

We know that the harmonic series diverges. We know that this divergent series times a constant, $\frac{1}{(1+Q)}$, still diverges. Now we have another series, $\sum_{n=1}^{\infty} \frac{1}{(1+nQ)}$, which is term by term *greater* than that divergent series—so it must diverge too. That is, if we assume that our series $\sum_{i=1}^{\infty} \frac{1}{p_i}$ converges (so that Q exists), the series $\sum_{n=1}^{\infty} \frac{1}{(1+nQ)}$ must *diverge*.

We will quickly show, however (again using our assumption), that $\sum_{n=1}^{\infty} \frac{1}{(1+nQ)}$ *can't* diverge: it must converge—and this will be the desired contradiction.

Why must $\sum_{n=1}^{\infty} \frac{1}{(1+nQ)}$ converge? Look first at all n for which $1 + nQ$ has only one factor among the primes which are greater than p_k. For *each* of those n's, $\frac{1}{1+nQ} = \frac{1}{p_v}$ for some prime $p_v > p_k$. So if we add together *all* such cases (even if there are infinitely many of them) $n_1, n_2, \ldots$, each with its separate $p_v, p_w, \ldots$, we'll get

$$\frac{1}{(1+n_1Q)} + \frac{1}{(1+n_2Q)} + \ldots = \frac{1}{p_v} + \frac{1}{p_w} + \ldots \leq \frac{1}{p_{k+1}} + \ldots < \frac{1}{2}$$

(by our definition).

Now look at all those n for which $1 + nQ$ has two prime factors, each greater than p_k. For each of these, $\frac{1}{(1+nQ)} = \frac{1}{p_v p_w}$ for some p_v and p_w. Adding *them* all together,

$$\frac{1}{(1+n_1 Q)} + \frac{1}{(1+n_2 Q)} + \ldots = \frac{1}{p_v p_w} + \frac{1}{p_x p_y} + \ldots \leq \left(\frac{1}{p_{k+1}} + \ldots\right)^2 < \left(\frac{1}{2}\right)^2$$

(for each $\frac{1}{p_v p_w}$ comes up somewhere in that squared term).

The n's for which $1 + nQ$ has three prime factors will give us

$$\frac{1}{(1+n_1 Q)} + \frac{1}{(1+n_2 Q)} + \ldots = \frac{1}{p_v p_w p_x} + \frac{1}{p_y p_z p_r} + \ldots \leq \left(\frac{1}{p_{k+1}} + \ldots\right)^3 < \left(\frac{1}{2}\right)^3$$

and so on.

If we now add up *all* these $\frac{1}{(1+nQ)}$, for every possible n, we will get some $< \frac{1}{2}$, some $< \left(\frac{1}{2}\right)^2$, others $< \left(\frac{1}{2}\right)^3$, still others $< \left(\frac{1}{2}\right)^4$, and so on, so that

$$\sum_{n=1}^{\infty} \frac{1}{1+nQ} < \sum_{t=1}^{\infty} \left(\frac{1}{2}\right)^t ,$$

and that right-hand sum, remember, *converges*. But a series with no negative terms and less than a convergent series must converge too—hence $\sum_{n=1}^{\infty} \frac{1}{(1+nQ)}$ both converges and diverges! This is the contradiction we sought, which proves that the sum of the reciprocals of the primes diverges.

This wonderfully acrobatic proof tells us something else: the number of primes is infinite. For were there only a finite number, this series would have to converge. It also reminds us not only that dealing with primes is always difficult, but that there is no problem that cannot be solved.

To Chapter Five

1. [to page 108] Why the circumcenter of a right triangle is the midpoint of the hypotenuse.

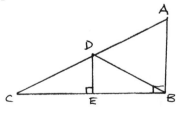

1. Drop a perpendicular from the midpoint, D, of the hypotenuse, meeting the opposite side at E.
2. Since $\triangle ABC \sim \triangle DEC$, and the ratio of similitude is 2:1, CE is half of CB—that is, CE = EB.

3. Draw DB. Then (by SAS) ΔDEB ≅ ΔDEC, so DB = DC.
4. Since CD = AD, D is equidistant from A, B, and C; hence, D is the circumcenter.

2. [to page 112] Why every triangle has a centroid.

<u>Theorem</u>: If in a triangle ABC two mass-balancing knife-edges AD and BE intersect at a point O, then every mass-balancing knife-edge passes through O, the centroid.

<u>Proof</u>:

1. Number the four regions into which these lines divide the triangle 1, 2, 3, and 4. We will now use these numbers to stand for the masses of their regions.

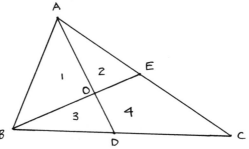

2. Since the median AD divides the triangle into two equal masses, 1 + 3 = 2 + 4.
3. Likewise, since BE is a median, 1 + 2 = 3 + 4.
4. So 1 + 2 = 3 + 4
      4 + 2 = 3 + 1
      ——————————
      1 − 4 = 4 − 1 ,
5. hence 1 = 4 and 2 = 3.
6. Now assume there is a mass-balancing knife-edge k that doesn't pass through O. Several situations are possible, of which we show one (proofs for the others are similar). Here k meets AB at Q, AD at a point P between A and O, AC at R.

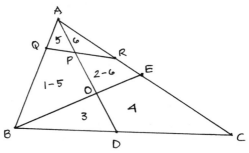

7. k creates regions 5 and 6, as shown, leaving region 1 – 5 and 2 – 6 below them.
8. Since k is mass-balancing, $5 + 6 = 1 - 5 + 3 + 2 - 6 + 4$.
9. After the appropriate subtractions and substitutions, from step 5, $2(5 + 6) = 2(1 + 2)$;
10. That is, $5 + 6 = 1 + 2$.
11. But this is impossible, since $5 + 6$ is a proper part of $1 + 2$. Therefore k passes through O.

<u>Corollary</u>: Since this theorem is true for every mass-balancing line, it is true for the median from C; hence the three medians coincide at the centroid O.

3. [to page 124] The nine-point circle, with the new tenth point on it.

This point P is where the three Euler lines (marked "e ΔKLC", etc.) of their relevant triangles concur.

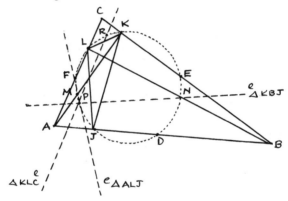

4. [to page 130] A proof that there is no shortest path in an obtuse triangle.

We first need an auxiliary theorem, often called a *lemma*. This one was devised by the ingenious Jim Tanton.

<u>Lemma</u> : If in ΔABC, with ∠B > 90°, there is a shortest path XYZ, then this path meets and leaves each side at equal angles.

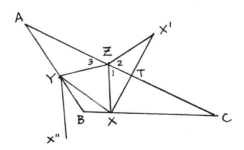

Proof:

1. Reflect XZ in AC to ZX' and draw XX', meeting AC at T.
2. $\triangle XZT \cong \triangle X'ZT$ (SSS), so $\angle 1 = \angle 2$.
3. Likewise reflect XY in AB to YX''. X''Y = XY (from congruent triangles, as in step 2).
4. If XYZX is shortest, then X''YZX' is shortest, and hence is a straight line, so $\angle 2 = \angle 3$.
5. Then by transitivity, $\angle 1 = \angle 3$, as desired.

The same argument applies to the other sides.

Theorem: There is no shortest path in an obtuse triangle.

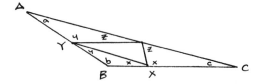

Proof:

1. For any candidate XYZ the angles (by our lemma) would be as lettered, and since
2.

$$\begin{array}{rrrr} \angle a + & \angle b + & \angle c & = 180° \\ \angle a + & \angle y + & \angle z & = 180° \\ \angle x + \angle b + \angle y & & & = 180° \\ \angle x + & & \angle c + \angle z & = 180° \\ \hline \end{array}$$

$2(\angle a + \angle x + \angle b + \angle y + \angle c + \angle z) = 720°$.
3. And since $\angle a + \angle b + \angle c = 180°$, $\angle x + \angle y + \angle z = 180°$.
4. But $\angle b > 90°$, so $\angle x + \angle y < 90°$, hence $\angle z > 90°$, and at vertex Z, $2(\angle z) > 180°$, which is impossible.

Hence there is no shortest path in an obtuse triangle.

5. [to page 130] The Fermat Point.

Here is that promised marvel of an example. The question itself seems innocent enough—a twin of the one at the end of Chapter Five. Is there a point P in a triangle such that the sum of its distances to the three vertices A, B, and C is minimal?

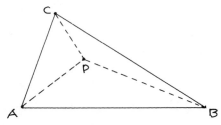

We want to minimize, that is, the sum of PA, PB, and PC, as anyone would who wanted to lay pipes most economically from a central pumping station to three consumers. Having learned our lesson, we will be more cautious this time and consider both acute and obtuse representative triangles.

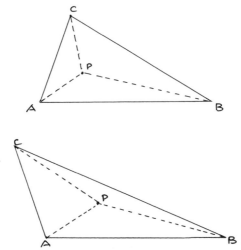

Shortest distance—straight line. Like an apprentice chess player, we are beginning to recognize the combinations. Remembering the gambit of false position, we let P be any point. But now what? Extending PA, PB, and PC their own lengths to new points X, Y, and Z gives a triangle similar to ∆ABC, which is simply our old problem drawn larger.

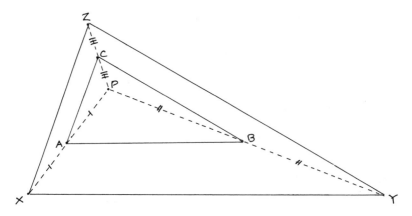

We need a zigzag line made up of segments equal to PA, PB, and PC, so that when we pull it taut, P will pop into the right place.

Perhaps thoughts like these went through Pierre de Fermat's mind when in the mid-seventeenth century he started work on this problem between his law cases. They may also have struck J. E. Hofmann three centuries later when he came up with a solution (Torricelli, to whom Fermat had sent the problem, devised a different one around 1640 or so, as have others since). But how can we possibly reconstruct the steps that led him to look at △APC *and rotate it 60° counterclockwise around A, to make the new △AP′C′?*

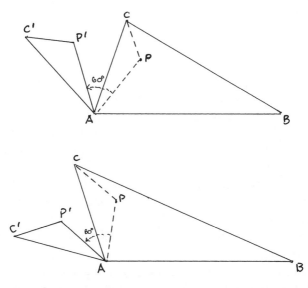

We'll try to retrieve his insight by thinking backwards while following his steps forward. He next draws P′P (ah—part of the broken line that will eventually be straightened) and notes that, since AP′ is just AP swivelled through 60°, AP′ = AP.

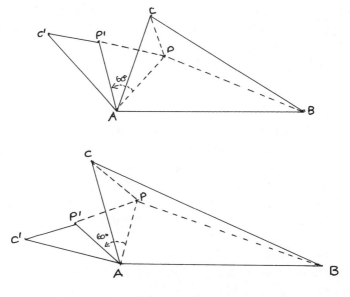

He knew the early theorem in Euclid—another result that Thales may first have seen—that if two sides of a triangle are equal, so are the angles opposite them:

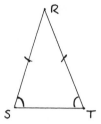

Euclid's proof involves a construction sufficiently difficult for beginners, and looking enough like a trestle bridge, to have earned the title *pons asinorum*: if you can cross it, you are a fool no longer. Six centuries later another Alexandrian, Pappus, came up with a proof so elegantly simple that you may wonder if it is a proof at all. Since SR = TR, ∠R = ∠R, and TR = SR, ΔSRT ≅ ΔTRS by SAS. Hence the corresponding parts are equal, among them ∠S = ∠T. Is this one-liner a joke or a proof that deserves to be in The Book? Notice that Pappus didn't prove a triangle congruent to itself—rather, two *different* triangles inhabiting the same body.

Here, then, since AP = AP′, the base angles of ΔAP′P are equal: ∠AP′P = ∠APP′. Each is therefore $\frac{(180°-60°)}{2}$ = 60°, and ΔAP′P is equilateral: all its angles and all its sides are equal. That means AP also equals P′P—and *now* we see what Hofmann was up to: that mysterious 60° rotation was just to achieve this equality! (How could the Formalist account possibly include this feel for the lay of the land, which leads to discovery?)

For Hofmann was interested in minimizing the sum of three lengths: AP + PB + CP. But AP = P′P, PB is equal to itself, and CP is equal to C′P′, since it was rotated into it by our 60° swing. So AP + PB + CP = PP′ +PB + C′P′, or, to put those in a more useful order,

$$C'P' + P'P + PB .$$

That crooked line, C′P′PB, is the one he wants to straighten. It will be straight when the angles at P′ and P are: that is, when ∠C′P′P and ∠P′PB are each 180°. We know *part* of each of these angles: ∠AP′P is 60° and so is ∠P′PA; so the remaining angles—∠C′P′A and ∠APB—each have to be 120°. But ∠C′P′A = ∠APC (the first is just the second rotated). This means that back in our original ΔABC, the angles around P made by PA, PB, and PC are each 120°!

The point P, then, from which the sum of the distances to the triangle's three vertices is least, is the point from which those three lines meet in pairs at 120° angles. You might think of the lines from P to vertices A, B, and C as being elastic cords, and as you move P around (with the cords lengthening and shortening) the angles around it change too; and you stop when they are all equal.

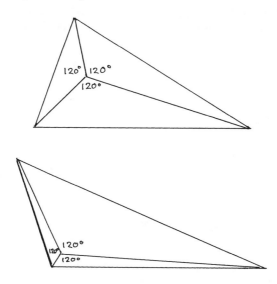

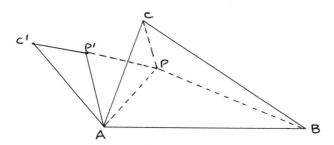

You may find this vaguely disappointing, because it doesn't really tell you how to find P. A beautiful solution, however, is hiding just around the diagram's corner. Look at it again

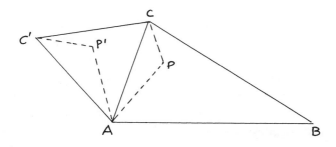

and draw in one more line: C′C.

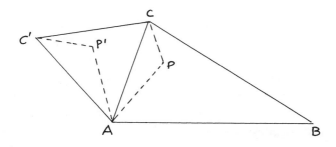

Why? Because wonderfully enough, $\triangle AC'C$ is equilateral too: AC' is just AC rotated, so $AC' = AC$, hence the angles $\angle AC'C$ and $\angle ACC'$ are equal—and since the hinge angle $\angle C'AC$ is $60°$, each of these is $\frac{(180°-60°)}{2}$ $= 60°$. We know we have our point P when C', P', P, and B are collinear; so if we simply *construct* an equilateral triangle on side AC, with new vertex C', and then connect C' and B, P must lie somewhere on this line.

And now we have an endgame like that in Fagnano's Problem: there was nothing special about side AC, so build an equilateral triangle on another side of $\triangle ABC$, such as BC, with vertex A':

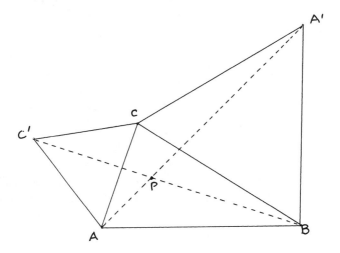

P must lie on $A'A$ also—so where $A'A$ and $C'B$ cross is the P we want, called the Fermat Point.

This very surprising simplification came from the previous hard work, the way a finished building emerges from its scaffolding. At least we were careful this time to test what we did on a representative obtuse as well as acute triangle—but were we careful enough, or has the Protean nature of things once again caught us off guard? Were those triangles *sufficiently* representative? Let's look at a very obtuse triangle, such as this:

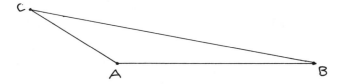

If we make our construction, we see that

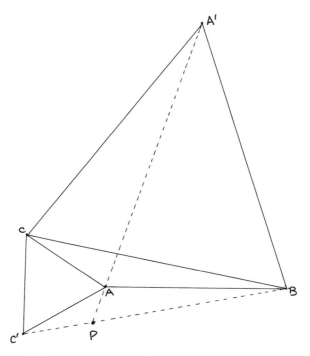

P has escaped! Our construction fails when an angle of our triangle is greater than or equal to 120°. When that happens, we must make do with the vertex A as the Fermat Point: one of those solutions that—like altitude AF in Fagnano's Problem or (back in Chapter One) calling a single dot a triangular or square number—pushes the envelope of definition.

To Chapter Six

1. [to page 140] Solving quadratic equations.

(a) Completing the square.

Our first method is called "completing the square". Starting from

$$0 = t^2 + t - 1 ,$$

just for the sake of neatness store all the terms with unknowns on one side of the equation, the known quantities on the other:

$$t^2 + t = 1 .$$

Were that left-hand side less messy—were it something squared—then we could take the square root of both sides and be just about done.

Well, says the voice of Alcibiades, make it so: add whatever is needed to $t^2 + t$ so that it *becomes* some expression squared. This is where algebra's stored up experience with factoring quadratics pays off. If you add $\frac{1}{4}$ *
to $t^2 + t$, you get $t^2 + t + \frac{1}{4}$, which is $\left(t + \frac{1}{2}\right)^2$.

But Alcibiades! You can't arbitrarily add something on to an expression—that changes its value!

You can, he answers, if you have an equation, as we do here. You keep the see-saw balanced by adding the same $\frac{1}{4}$ to the other side: the mathematician's twitchiness about asymmetry once more soothed.

$$t^2 + t + \frac{1}{4} = 1 + \frac{1}{4} \,.$$

Our original equation $t^2 + t = 1$ has now turned into the equivalent

$$\left(t + \frac{1}{2}\right)^2 = \frac{5}{4} \,.$$

Now we can take the square root of both sides (since that was the reason for all these gymnastics) and we'll consider only the positive square root, since lengths can't be negative. We get

$$t + \frac{1}{2} = \sqrt{\frac{5}{4}}$$

which simplifies to

$$t + \frac{1}{2} = \frac{\sqrt{5}}{2} \,.$$

Hence

$$t = -\frac{1}{2} + \frac{\sqrt{5}}{2} \,.$$

More simply,

$$t = \frac{-1 + \sqrt{5}}{2}$$

or, if you like your numbers arranged so that you can see in order the geometric operations that will happen to them:

$$t = \frac{\sqrt{5} - 1}{2} \,.$$

---

*How did the algebraist come up with 1/4? He needed some number w so that $(t + w)^2 = t^2 + 2wt + w^2$ would in fact yield $t^2 + t +$ some number: that is, 2w must be 1 (the coefficient of t), so w must be $\frac{1}{2}$; and $\left(\frac{1}{2}\right)^2 = \frac{1}{4}$. A choice instance of moving forward by thinking backward.

(b) The Quadratic Formula, which formalizes the technique of completing the square.

To solve for x in $ax^2 + bx + c = 0$, we have to keep freeing x from its various bonds (other numbers added to and multiplied by it, and being squared).

Begin by subtracting the constant c from both sides:

$$ax^2 + bx = -c.$$

To keep things as simple as possible, divide by a:

$$x^2 + \left(\frac{b}{a}\right) \cdot x = \frac{-c}{a}.$$

Now ask yourself (as in the first approach) what needs to be added to the left side of this equation in order to turn it into a perfect square. This is the key piece of the puzzle, because finding it will allow us to take the square root of both sides and so free x from its exponent. Those who found it, like Bhāskara (in twelfth-century India), are immortalized in the mathematical Pantheon. With some trial and error—or insight— you come up with $\frac{b^2}{4a^2}$: for then

$$x^2 + \left(\frac{b}{a}\right) \cdot x + \frac{b^2}{4a^2} = \left(x + \frac{b}{2a}\right)^2.$$

Adding $\frac{b^2}{4a^2}$, then, to both sides, we have

$$\left(x + \frac{b}{2a}\right)^2 = \frac{b^2}{4a^2} - \frac{c}{a},$$

and putting the right-hand side over the common denominator of $4a^2$,

$$\left(x + \frac{b}{2a}\right)^2 = \frac{b^2 - 4ac}{4a^2}.$$

Now we can take the square root of both sides:

$$x + \frac{b}{2a} = \frac{\pm\sqrt{b^2 - 4ac}}{2a}$$

so that

$$x = \frac{-b \pm \sqrt{b^2 - 4ac}}{2a}.$$

If you put the coefficients a, b, and c of any quadratic equation $ax^2 + bx + c = 0$ into this Quadratic Formula, the equation's two roots will come out.

2. [to page 150] Why $\sqrt{3}$ is not in F_1.

Suppose there *were* rationals $\frac{a}{b}$ and $\frac{c}{d}$ such that

$$\frac{a\sqrt{2}}{b} = \frac{c\sqrt{3}}{d}.$$

Remembering what we did in Chapter One, we square both sides and rearrange, getting

$$2(ad)^2 = 3(bc)^2.$$

Since ad is some natural number—call it r—and bc a natural number s, this simplifies to

$$2r^2 = 3s^2.$$

If we break r and s down to their ultimate, prime factors, each of these primes will appear twice on each side (since r and s are each squared); so on the left-hand side, 2 will appear an odd number of times (possible pairs from the factors of r^2, and that solitary initial 2) but an even number of times (none or pairs) on the right. Divide both sides by 2 as many times as you can, and you'll be left with one surviving 2 on one side of the equation or the other—which will make its side even and the other side odd. But an odd number can't equal an even number, so we never had a true equation. This proof by contradiction shows that $\sqrt{3}$ cannot belong to F_1's society.

3. [to page 165] On Hermes's work.

The *Diarium*, in which Hermes wrote up his ten years of work, comes with peculiar puzzles. Anyone undertaking to show just how to construct the 65,537-gon would of course know that Gauss had proved it to be constructible (since $65,537 = 2^{2^4} + 1$, a Fermat prime)—and after constructibility, actual construction is something of an anticlimax. Why did Hermes spend a decade on what could only be uninformative details?

One possibility is that the devil in these details really does conceal something interesting. Hermes may (as Paddy Patterson at Göttingen conjectures) have been using the many different ways of expressing 65,537 as a sum of two squares, in order to determine the square roots he needed. If this were so, however, he would have had no need to express his square roots numerically—certainly not to ten or more decimal places, as he did.

Another possible answer might follow from his title page, which has a drawing showing the segment midway along (between vertices 32,768 and 32,769) and about 11.3 cm ($\approx$ 4.5 inches) long, and states that to have the polygon's side-lengths this size would require it to be inscribed in a unit circle of radius 1,168.32 meters ($\approx$ 3,797 feet or about seven tenths of a mile). Was Hermes planning to have his model actually constructed—as a great monument, perhaps, to Gauss? We know that Gauss had hoped (in vain) that the 17-gon would be inscribed on his tombstone. Did Hermes wish to make up for such a slight? He lived at a time when making physical models of geometrical objects was commonly carried out, with great skill and patience. As late as 1951 a book of instructions for making immensely complicated models of polyhedra was published, and a wire model of one with 720 faces and 1200 edges had been built. The virtue of such models lies in their giving romantic reality to imagination, though their particularity draws thought away from its proper, relational, realm.

Yet if making a physical model had been Hermes's intent, why did he carry his calculations out so far? Three decimal places rather than ten would have been more than wood or wire could tolerate.

His title page contains another clue, however. There at the hub of the circumcircle is his dedication: to the Manes (the Roman equivalent of soul) of Richelot—the man who, with Schwendenwein, had forty years before calculated how to construct the 257-gon. Had he set Hermes this doctoral task, thinking it commensurate with his dogged skills—and had the student stuck to it through a decade of work, in homage to the spirit of his master that now lay entangled within it? Read so, the story rings with not Roman but Wilhelmine virtue.*

A more cynical view can be found in this curious passage by the English number theorist J. E. Littlewood (from the "Cross-Purposes" section of *Littlewood's Miscellany*, ed. B. Bollobás, Cambridge University Press, 1986, p. 60):

> A too-persistent research student drove his supervisor to say 'Go away and work out the construction for a regular polygon of 65537 sides.' The student returned 20 years later with a construction (deposited in the Archives at Göttingen).

---

*Under the Prussians, discipline was so prized that even soldiers killed in battle were reputed to snap to attention at the trumpet's call: *Kadavergehorsamkeit*, "corpse obedience", described this dedication to duty.

Some credence is given this version by Littlewood's near contemporaneity with the event, but diminished by his parodic intent and what may have been the different *bien entendus* of English academic life.

Might the answer not lie at some point inside the triangle whose vertices are the romance of numbers, of obligation, and of obsession? Hermes wouldn't have been alone in lowering himself down into an irrational crevice in the number line, only to be drawn ever deeper, by the promise of some conclusive revelation only a decimal place further on. Was this promise not the devil who dances our souls away in detail?

To Chapter Seven

1. [to page 194] The three complex roots of 1.

In their street clothes, algebraists calculate with patience and accuracy. When they change from Clark Kent in a phone booth, they emerge eager to spark an insight through equating very different-seeming expressions. We will need both of their embodiments in order to find the three cube roots of 1: that is, the real numbers a and b such that $(a + bi)^3 = 1 + 0i$. Spelling this out,

$$a^3 + 3a^2bi - 3ab^2 - b^3i = 1 + 0i .$$

Equating real part with real part, imaginary with imaginary, we find that

$$a^3 - 3ab^2 = 1$$

and

$$3a^2b - b^3 = 0 .$$

Since 1 (that's $1 + 0i$) is, after all, a cube root of 1, $a = 1$ and $b = 0$ must be one of the three solutions. To find the other two we may therefore assume $b \neq 0$. That lets us divide by b in this last equation, giving us

$$3a^2 - b^2 = 0$$

or

$$b^2 = 3a^2 .$$

Taking the square root of both sides,

$$b = \pm a\sqrt{3}.$$

Now substitute $b = +a\sqrt{3}$ into the first equation

$$a^3 - 3ab^2 = 1,$$

to get $a^3 - 9a^3 = 1$; in other words,

$$-8a^3 = 1.$$

This gives us

$$a = \sqrt[3]{\left(\frac{-1}{8}\right)},$$

so

$$a = -\frac{1}{2}.$$

Hence

$$b = a\sqrt{3} = \frac{-\sqrt{3}}{2},$$

giving us the second cube root of 1:

$$-\frac{1}{2} - \frac{\sqrt{3}}{2}i.$$

Faster than a speeding bullet, the other possible choice for b, $-a\sqrt{3}$, yields the third root,

$$-\frac{1}{2} + \frac{\sqrt{3}}{2}i.$$

2. [to page 199] The non-constructibility of the heptagon.

You may have felt short-changed when, in the previous chapter, we only stated but didn't prove Gauss's conclusion about what kinds of polygon can be constructed. Let's atone for that now by showing why the heptagon, at least, can't be: paradoxically, its emergence far away on the complex plane both hints that it might be constructible and proves that it isn't.

The strategy is this. $x^7 = 1$, or $x^7 - 1 = 0$, is the equation for the seven roots of unity, and hence (on the complex plane) for the vertices of the heptagon, as we saw at the end of the chapter.

By a series of deft moves we'll make this equation yield a cubic involving the cosine of the angle ($\phi = \frac{2\pi}{7}$ radians) at the center of each of the heptagon's pie-slices. The length $\cos \phi$ is involved with the length of the heptagon's side, so if *it* can't be constructed, neither can that side:

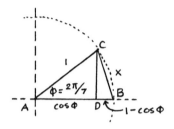

If x could be constructed, so could $\triangle ABC$, whose other sides are the radii (length 1) of the circumscribed circle. So, therefore, could the perpendicular CD to AB, and thus AD = cos ϕ could be constructed too. If x could be constructed, then cos ϕ could be—but cos ϕ cannot, so neither can x.

This is what we will show, by discovering that the cubic which follows from the heptagon's equation has no roots in a square root extension field, where (as we know from Chapter Six) the only constructible lengths lie.

Our order of operations will be first to find out what sort of cubics *don't* have roots in square root extension fields—and only then to reduce the heptagon's equation to a cubic, and see that it is of this sort.

(a) Cubics and their roots.

Think of yourself as a traveler in a medieval landscape. There are six castles ahead, guarding the route. It won't take trials of strength to get past them, but resoluteness in the face of their equations—which only look like portcullises: they turn out to be drawbridges. A morning sort of optimism will help too, since we tend to tolerate only a few unaccustomed turns of thought in a day. Each leaves a residue of discomfort: did it really work? Did I really understand it? Too many such, and a sense of the whole, along with self-confidence, topples (doesn't the solution to the riddle of intuition lie here? We take as intuitive whatever *use* has made so familiar that we casually apply it to other ends).

The first of these six castles contains *The Factor Theorem*, which Descartes came up with in 1637. We know how helpful it is to break down numbers into their prime factors (such as $6 = 2 \cdot 3$) and so can imagine that it must be equally valuable to factor polynomials.

If t is a root of a polynomial, f(x), by definition

$$f(t) = 0 \, .$$

We should like to prove that $(x - t)$ is a factor of f(x)—that is, that $(x - t)$ *divides* f(x), with no remainder.

Look at the worst case. You divide $(x - t)$ into f(x) and get a quotient, Q, and a remainder, R. When you divide 7 into 45, for example, the quotient is 6 and the remainder is 3. Notice that you can write

$$45 = 7 \cdot 6 + 3.$$

In the same way, you can write

$$f(x) = (x - t) \cdot Q + R \, .$$

When x = t, you get

$$f(t) = (t - t) \cdot Q + R \, .$$

But f(t) = 0 (since t is a root), and $(t - t) = 0$, so

$$0 = 0 \cdot Q + R \, ,$$

which means R = 0.

There is no remainder, so $(x - t)$ is indeed a factor of f(x).

Descartes is also the keeper of the second castle, which contains his *Rational Roots Theorem*: if a polynomial f(x) has a *rational* root, this theorem gives us a list of the possible candidates.

Early in our algebra careers we learn that given, say,

$$(2x - 3)(x + 4) = 2x^2 + 5x - 12,$$

i.e., that $(2x - 3)$ and $(x + 4)$ are the factors of $2x^2 + 5x - 12$—then, if

$$2x^2 + 5x - 12 = 0,$$

it must be true that $(2x - 3)(x + 4) = 0$.

Since the only way to have the product of two factors equal to zero is to have at least one of them *be* zero,

$$\text{either} \quad 2x - 3 = 0, \quad \text{so } x = \frac{3}{2}$$

$$\text{or} \qquad x + 4 = 0, \quad \text{so } x = -4 \, ,$$

and we have solved our polynomial. Note the parallel to the Factor Theorem: there Descartes saw that if t is a root, $(x - t)$ is a factor. Here we discover that if $(x - t)$ is a factor, t is a root.

The only difficult part of this technique for solving polynomials by factoring is finding the factors—but a moment's thought shows us that the possible *candidates* for factors are determined by the polynomial itself. Look again at

$$2x^2 + 5x - 12 = 0$$

and set up dummy parentheses to signify its factors:

$$(cx + d)(ex + g) = 0.$$

What could c and e possibly be? They would have to be integers which multiply together to make 2—so can be only ±1 or ±2. Similarly, d and g have to multiply together to make 12, so can be only (±) 1, 2, 3, 4, 6, and 12.

Now, take *any* polynomial

$$f(x) = ax^n + bx^{n-1} + \ldots + j$$

and set up even *one* dummy factor

$$(mx + n) ;$$

m would have to be a factor of a,
n would have to be a factor of j.

f(x) might, of course, have *no* factors, but *if* $(mx + n)$ is a factor of f(x), then $\frac{-n}{m}$, a rational number, is a root.

So we can say that *any* rational root has to have a denominator which is a factor of the coefficient of the highest term of the polynomial, and a numerator which divides the polynomial's constant term.

Since both the Factor Theorem and the Rational Roots Theorem are true for any polynomial, they are certainly true for the cubics we are interested in.

Girolamo Cardano—one of the strangest figures in all of mathematics—lives in the third castle (he flickered briefly past us on page 170). A century before Descartes, this thoroughly Renaissance man boasted, cringed, calculated, cheated, invented, and lied his way through Italian life. His uncle, daughter-in-law, and protégé were all poisoned, his son beheaded, and he himself thrown in prison for blasphemously casting the horoscope of Christ. He cured Scotland's Archbishop of asthma by

Girolamo Cardano (1501–1576), a man whose modesty modestly made way for his self-confessed excellence.

sheer reason and wrote seven thousand pages on everything from navigation to the black arts. You may take the man's curious measure from this passage in his autobiography, on the marvel of movable type: "What lack we yet unless it be the taking of Heaven by storm! Oh, the madness of men to give heed to vanity rather than the fundamental things of life! Oh, what arrogant poverty of intellectual humility not to be moved to wonder!"

Cardano was the first to reckon odds; he worked on the construction of the pentagon, on cubic and quartic equations—and found that if $x^3 + bx^2 + cx + d = 0$, then the sum of the polynomial's three roots will be equal to $-b$. This is the result we will need, and here is how he got it.

Let's call the three roots of this cubic t, u, and v. By the factor theorem, we now know that $(x - t)$, $(x - u)$, and $(x - v)$ will each be a factor of the polynomial. They must also be its *only* factors, because a cubic has exactly three roots. There can't be a *constant* multiplier, because our polynomial starts "x^3", not "ax^3". This means that

$$(x - t)\,(x - u)\,(x - v) = x^3 + bx^2 + cx + d\,.$$

It may seem like twiddling your thumbs while waiting for inspiration, but let's multiply out the left-hand side:

$$x^3 - (t + u + v)\,x^2 + (tu + tv + uv)\,x - tuv = x^3 + bx^2 + cx + d\,.$$

This can only mean that the two different-looking coefficients of x^2 *are the same*:

$$-(t + u + v) = b\,,$$

or

$$t + u + v = -b$$

and that is just what Cardano proved, between bouts of necromancy and vituperation. One more instance, then, of the algebraist giving depth to an object by looking at it from two different standpoints.

Appendix

At the fourth castle we take a refreshing pause. If a complex number x is trigonometrically $\cos \phi + i \sin \phi$, what will $\frac{1}{x}$, that is: $\frac{1}{(\cos \phi + i \sin \phi)}$, look like? We'll use once again the conjugacy tactic of pages 150 and 170:

$$\frac{1}{(\cos \phi + i \sin \phi)} \cdot \frac{(\cos \phi - i \sin \phi)}{(\cos \phi - i \sin \phi)} = \frac{(\cos \phi - i \sin \phi)}{(\cos^2 \phi + \sin^2 \phi)} .$$

But $\cos^2 \phi + \sin^2 \phi = 1$ (as you saw on page 183), so

$$\frac{1}{x} = \cos \phi - i \sin \phi .$$

Graphically:

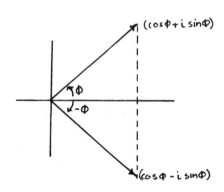

Each of our four brief visits abbreviates long swathes of time when people puzzled over what we now take for granted, just as each of our words condenses receding landscapes of thought. At least in mathematics, ontogeny recapitulates phylogeny.

Take, for example, a neat technique which will help in the castle ahead. Its inventor is no more remembered than whoever first thought of putting in a little cream and reducing it when making a *beurre blanc*, to keep the sauce from breaking down. Here, if you have a polynomial equation with rational coefficients, such as

$$ax^3 + rx^2 + sx + v = 0,$$

you can divide the equation by a without changing its nature and still leave its coefficients rational:

$$x^3 + \left(\frac{r}{a}\right)x^2 + \left(\frac{s}{a}\right)x + \left(\frac{v}{a}\right) = 0 .$$

For ease, let $\frac{r}{a} = b$, $\frac{s}{a} = c$, and $\frac{v}{a} = d$, giving us

$$x^3 + bx^2 + cx + d = 0 .$$

As you've seen, it's just that much easier to handle polynomials whose highest coefficient is 1—our gratitude, then, for such ball-bearing ingenuities on which grand enterprises glide.

The fifth castle is larger than the others: it houses an insight into the roots of a cubic equation with rational coefficients, like $x^3 + bx^2 + cx + d = 0$. For we want to show that the paired forms you have already seen several times reappear here: namely, if $p + q\sqrt{w}$ should happen to be one root of this polynomial, then $p - q\sqrt{w}$ would have to be another ($\sqrt{w}$ first appears in some square root extension field, F_k, of the rationals, F; but p, q, and w are in F_{k-1}).

This pairing of roots seems a likely proposition. Since a cubic has three roots, if we had found one that was rational, r, then $(x - r)$ would be a factor of the polynomial, and its paired factor would have to be a quadratic:

$$p(x) = (x - r)(x^2 + bx + c).$$

The other two roots would then come out of the quadratic formula:

$$x = \frac{-b \pm \sqrt{(b^2 - 4ac)}}{2a},$$

so that if one root was $p + q\sqrt{w}$, its mate would have the paired sign: $p - q\sqrt{w}$. Unfortunately we haven't yet found one of the roots so we don't know that it is a rational number, r.

Or you might think that since we now know that the sum of the three roots is equal to the rational number –b, the only way to rid ourselves of the irrational $q\sqrt{w}$ in one root must be to add $-q\sqrt{w}$ in another. This is very reasonable (and turns out to be true)—but we just don't know enough about irrationals at this point to be certain that some quite different irrational added to $q\sqrt{w}$ might not yield a rational sum (after all, the two irrationals

$$0.10110111011110\ldots$$

$$0.01001000100001\ldots$$

add up to the rational $0.111\ldots = \frac{1}{9}$).

Not proven, but hunch rallies our hopes high enough to make our way though the delicate negotiations ahead. We'll simply find out what $f(p + q\sqrt{w})$ is, and follow the consequences. Then we'll go back and replace x by $p - q\sqrt{w}$ and follow *its* consequences: $p - q\sqrt{w}$ will also turn out to be a root (and as a final flourish, a *different* root from $p + q\sqrt{w}$).

Appendix

Here we go. Putting $p + q\sqrt{w}$ for x in

$$x^3 + bx^2 + cx + d = 0$$

gives us

$$(p + q\sqrt{w})^3 + b(p + q\sqrt{w})^2 + c(p + q\sqrt{w}) + d = 0 .$$

We must spell this all out in order to regroup and see what we have:

$$p^3 + 3p^2q\sqrt{w} + 3pq^2w + q^3w\sqrt{w} + bp^2 + 2bpq\sqrt{w} + bq^2w + cp + cq\sqrt{w} + d = 0 .$$

Rearranging and giving the names m and n to our clusters,

$$\underbrace{(p^3 + 3pq^2w + bp^2 + bq^2w + cp + d)}_{m} + \underbrace{(3p^2q + q^3w + 2bpq + cq)}_{n}\sqrt{w} = 0$$

we arrive at

$$m + n\sqrt{w} = 0 ,$$

where m and n both belong to F_{k-1}.

Both m and n must be 0. Why? Because if n weren't, we could divide by it, so

$$\frac{m}{n} + \sqrt{w} = 0 ,$$

or

$$\sqrt{w} = \frac{-m}{n} ,$$

which is impossible: $\frac{-m}{n}$ is rational and $\sqrt{w}$ isn't (remember that w belonged to F_{k-1} but $\sqrt{w}$ didn't: it is in its own square root extension field). So we can't divide by n: hence n = 0. This means that

$$0 = m + 0\sqrt{w} = m ,$$

that is, m = 0 too.

If we now replace x by $p - q\sqrt{w}$, we will get the same polypedalian creature as before (what might not Nichomachus have called it?), but with minus signs instead of plus signs wherever q appears to an odd power, which will be in the coefficient of $\sqrt{w}$.

$$\underbrace{(p^3 + 3pq^2w + bp^2 + bq^2w + cp + d)}_{m} + \underbrace{(-3p^2q - q^3w - 2bpq - cq)}_{-n}\sqrt{w}$$

Our polynomial turns into $m - n\sqrt{w}$, and since these are the same m and n which we just proved were 0,

$$m - n\sqrt{w} = 0,$$

hence $p - q\sqrt{w}$ is a root of the polynomial too—and a different root at that: for if $p + q\sqrt{w} = p - q\sqrt{w}$, then $2q\sqrt{w} = 0$, which would make $q = 0$, so the original root $p + q\sqrt{w}$ would have been just p, which is in F_{k-1} (contrary to our assumption that this root first appeared in F_k).

We have come to the last castle, which guards the pass to the heptagon. In it is the secret of which cubics lack roots in square root extension fields of the rationals. The secret (whose clues lay in the previous castles) is this: if a cubic equation with rational coefficients has *no rational roots*, then in fact *none* of its roots lie in any square root extension field, F_k, of the rationals. For assume that b, c, and d are rational, and that one of the roots of

$$x^3 + bx^2 + cx + d = 0$$

does indeed appear for the first time in some F_k, and so looks like $p + q\sqrt{w}$. The fifth castle's guardian assures us that $p - q\sqrt{w}$ is also a root; the Fundamental Theorem of Algebra (page 169) that there must be a third root—call it v; and Cardano then exclaims that these three roots must add up to –b:

$$p + q\sqrt{w} + p - q\sqrt{w} + v = -b,$$

in other words,

$$2p + v = -b$$

or

$$v = -b - 2p.$$

But $-b - 2p$ appears in F_{k-1} (since b is rational and hence is in F and every square root extension field of F, and p was explicitly stated on page 297 to be in F_{k-1}), contradicting our assumption that no root of our equation appears until F_k. The sixth castle has yielded up its secret and we are through the pass. Now we need only reduce the heptagon's

Appendix

equation to a cubic and find that it has rational coefficients but no rational roots.

(b) Reducing the heptagon's equation.

Watch how all the parts now click into place. $x^7 - 1 = 0$ is indeed an equation with rational coefficients—but it isn't a cubic. A sequence of really artful moves (Renaissance born, in the spirit of projecting the vast down to human scale) will draw a cubic out of it.

We know that 1 is a root of this equation, so by our first discovery, $(x - 1)$ is a factor of $x^7 - 1$. Since it is a factor, we can find the other factor by dividing:

$$\frac{x^7 - 1}{x - 1} = x^6 + x^5 + x^4 + x^3 + x^2 + x + 1$$

(you can verify this by multiplying the right-hand side by $x - 1$).

The six other roots of $x^7 - 1$ will come from

$$x^6 + x^5 + x^4 + x^3 + x^2 + x + 1 = 0 .$$

The need for a cubic, along with informed tinkering, leads to the next step: dividing both sides of this new equation by x^3 (secure in the knowledge that $x \neq 0$: since if it were, the equation above would tell us that $1 = 0$). This gives us:

$$x^3 + x^2 + x + 1 + \left(\frac{1}{x}\right) + \left(\frac{1}{x^2}\right) + \left(\frac{1}{x^3}\right) = 0 .$$

The first four terms are comfortable, the last three disturbing. To deal with them, here is a cunning but legitimate rearrangement:

$$\left[x^3 + \left(\frac{1}{x^3}\right)\right] + \left[x^2 + \left(\frac{1}{x^2}\right)\right] + \left[x + \left(\frac{1}{x}\right)\right] + 1 = 0 .$$

Why do this? Because—and this was an innovation as slick as the curve ball—this equation can in turn be transformed into

$$\left[x + \left(\frac{1}{x}\right)\right]^3 - 3\left[x + \left(\frac{1}{x}\right)\right] + \left[x + \left(\frac{1}{x}\right)\right]^2 - 2 + \left[x + \left(\frac{1}{x}\right)\right] + 1 = 0 .$$

If you expand the terms that need expansion, you will see that the cancellations and sums of like terms return us to the previous equation.

Easy to check—but how did anyone ever think of doing it? One of the trade secrets of mathematics is to add zero to an expression in the useful form of what-you-want plus its additive inverse.

Simplify and rearrange once more:

$$\left[x+\left(\frac{1}{x}\right)\right]^3+\left[x+\left(\frac{1}{x}\right)\right]^2-2\left[x+\left(\frac{1}{x}\right)\right]-1=0 .$$

Still too bulky for comfort, but squint your eyes to see a simple cubic with rational coefficients, disguised by a complicated variable. All we really have here is

$$y^3 + y^2 - 2y - 1 = 0$$

where $y = \left[x+\left(\frac{1}{x}\right)\right]$.

In all of these contortions, we don't want to lose sight of what matters: the latent heptagon. We have just arrived at $y = \left[x+\left(\frac{1}{x}\right)\right]$: that x is still a vertex of our heptagon and that x in trigonometric form is still $\cos\phi + i\sin\phi$.

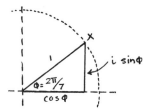

We brought along from the fourth castle the little plaything

$$\frac{1}{x} = \cos\phi - i\sin\phi ,$$

so

$$y = \left[x+\left(\frac{1}{x}\right)\right] = (\cos\phi + i\sin\phi) + (\cos\phi - i\sin\phi) = 2\cos\phi .$$

If we can't construct y, then we can't construct the length $2\cos\phi$, and therefore can't construct $\cos\phi$ through bisecting, and so cannot construct the heptagon with Euclidean tools.

So it all comes down to showing that

$$y^3 + y^2 - 2y - 1 = 0$$

has no rational root.

Suppose it had. Were $\frac{r}{s}$ (a rational in lowest terms) a root of this equation, then, from the second castle, r must be a factor of –1 and s of 1: in other words, r must be 1 or –1 and so must s, so that $\frac{r}{s} = 1$ or $\frac{r}{s} = -1$. Those are the *only possible* rational roots of our latest equation.

Yet if you try each, you find

$$f(1) = 1^3 + 1^2 - 2 \cdot 1 - 1 = -1; \text{ which isn't } 0,$$

$$f(-1) = (-1)^3 + (-1)^2 - 2 \cdot (-1) - 1 = 1; \text{ which also isn't } 0.$$

Neither is a root; hence, *our equation has no rational root*—and on this slender outcome of a long campaign, the battle is won: the heptagon cannot be constructed by straightedge and compass. As Wellington said of Waterloo: "It has been a damn'd nice thing—the nearest-run thing you ever saw in your life. . ."

Afterthoughts in the tent, or You Can't Get There from Here. Why do some parts of mathematics need so much more work than others? Why was our route to this result so devious (all those facts about cubics to capture a general result), when some that seem equally inaccessible turn out to be next door, and others that ought to be neighbors have still to be reached? We hardly yet grasp the lay of the land. The long frontier of mathematics expands like the Roman Empire's, through a shapeless unknown. The Teutoberg Forest may be just over the horizon.

To Chapter Eight

1. [to page 215] Finding invariants on the projective plane.

To gauge just how bad things are on the projective plane, notice that we can even project any three points on one line onto *any* three points of your choice on another! Let A, B, and C be arbitrary points on a line ℓ

and A′, B′, C′ equally arbitrary points on another line m:

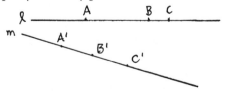

With the aid once again of a subtle diagonal, a chain of two perspectivities will do the work. Draw this diagonal n from A′ to C, then construct the lines A′A and B′B, which will meet at some point O (this is projective geometry: any two lines must meet):

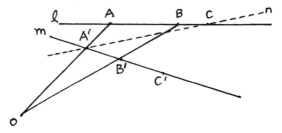

That line, OB′B, must also meet the diagonal line n somewhere: call it B″.

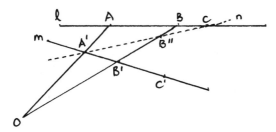

Next construct the lines OC and CC′. Line CC′ will meet BB′ at a point P:

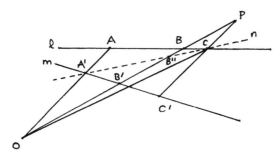

Finally, draw PA′.

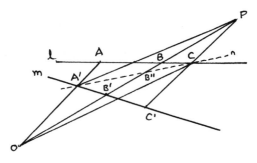

With O as our first center of perspectivity, the points A, B, and C on ℓ are sent to A′, B″, and C, respectively, on n. Let's write—as if O were a function—O(ABC) = A′B″C. Now with P as the second center of perspectivity, the points A′, B″, and C on n go to A′, B′, and C′ on m: P(A′B″C) = A′B′C′. Do O, then P, and we send A, B, and C to A′, B′, and C′—as desired. We could also write: P(O(A, B, C)) = A′, B′, C′.

Clever, but dreadful. The line n stands, like Schumann, between form and shadow, and only confirms our suspicion that nothing on the projective plane is stable. But to indulge in mathematics is to have faith in pattern: faith that with enough—or the right kind of—probing, fixity will emerge from change.

The very nature of perspectivity means there can be no similar triangles on the projective plane: angles won't stay the same under projection, nor lengths, so there can be no question of equal ratios. But take, as Melville suggests in *Moby Dick*, a deeper cut. Let's look not at three but at an arbitrary *four* points on a line ℓ. Send ℓ to any other line m by a perspectivity from some point O, $\ell \overset{O}{\wedge} m$, so that A, B, C, and D go to four points A′, B′, C′, and D′ on m: O(A, B, C, D) = A′, B′, C′, D′.

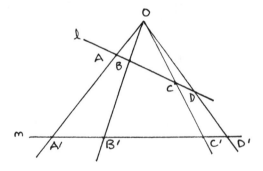

Is anything invariant here, no matter how far apart ℓ and m may be, or how differently inclined to one another? As long ago as the fourth century A.D., Pappus of Alexandria (whom you saw fleetingly in Chapter Five) uncovered a buried relation—in the context, however, of Euclidean geometry, with its native angles, lengths, and areas. Let's descend from the projective plane to the Euclidean to see what he found—though we will put it (as did the nineteenth-century German mathematician Augustus Ferdinand Möbius) in terms of the trigonometry we mastered in Chapter Seven.

Extract from our diagram ΔOAC, for example:

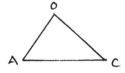

Its area is half the base, CA, times the altitude h to that base from O.

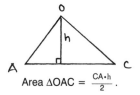

Area $\Delta OAC = \frac{CA \cdot h}{2}$.

We would, of course, have gotten the same area had we chosen OA as the base, with the altitude k to it from C:

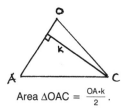

Area $\Delta OAC = \frac{OA \cdot k}{2}$.

Trigonometry reminds us that the sine of an angle is the ratio of the lengths: $\frac{\text{side opposite}}{\text{side adjacent}}$ in a right triangle, so that here $\sin \angle AOC = \frac{k}{OC}$. This means that $k = OC \cdot \sin \angle AOC$, so that we could if we wanted rewrite the triangle's area as

$$\text{Area } \Delta OAC = \frac{OA \cdot OC \cdot \sin \angle AOC}{2}.$$

The two expressions for area must be equal:

$$\frac{CA \cdot h}{2} = \frac{OA \cdot OC \cdot \sin \angle AOC}{2}.$$

From this equation it follows that

$$CA \cdot h = OA \cdot OC \cdot \sin \angle AOC.$$

Go through precisely the same maneuvers for ΔOCB to get

$$CB \cdot h = OB \cdot OC \cdot \sin \angle COB.$$

Should we then care to find the ratio of CA to CB, we would get

$$\frac{CA}{CB} = \frac{CA \cdot h}{CB \cdot h}$$

$$= \frac{OA \cdot OC \cdot \sin \angle AOC}{OB \cdot OC \cdot \sin \angle COB}$$

$$= \frac{OA \cdot \sin \angle AOC}{OB \cdot \sin \angle COB}.$$

Past adventures will give you confidence that we are winding our way into the heart of a labyrinth for the sake of coming out enriched on the other side.

Repeat these operations for $\triangle ODA$ and $\triangle ODB$:

$$\frac{DA \cdot h}{2} = \text{Area of } \triangle ODA = \frac{OA \cdot OD \cdot \sin \angle DOA}{2}$$

and

$$\frac{DB \cdot h}{2} = \text{Area of } \triangle ODB = \frac{OB \cdot OD \cdot \sin \angle DOB}{2} .$$

So that—as before—

$$\frac{DA}{DB} = \frac{OA \cdot OD \cdot \sin \angle DOA}{OB \cdot OD \cdot \sin \angle DOB}$$

$$= \frac{OA \cdot \sin \angle DOA}{OB \cdot \sin \angle DOB} .$$

No gap in the hedge appears—until we take the *ratio* of our two ratios:

$$\frac{\dfrac{CA}{CB}}{\dfrac{DA}{DB}} = \frac{\dfrac{OA \cdot \sin \angle AOC}{OB \cdot \sin \angle COB}}{\dfrac{OA \cdot \sin \angle DOA}{OB \cdot \sin \angle DOB}} = \frac{\dfrac{\sin \angle AOC}{\sin \angle COB}}{\dfrac{\sin \angle DOA}{\sin \angle DOB}} .$$

This double abstraction has vaporized the lengths, leaving only the sines of angles behind—but all these angles at O were determined by the original four points on line ℓ: they won't change no matter what line m we draw, and neither will their sines. That ratio of ratios, or *cross ratio*, as it is called, is a constant! It is the same whether we look at A, B, C, D on ℓ or at A′, B′, C′, D′ on m, hence

$$\frac{\dfrac{CA}{CB}}{\dfrac{DA}{DB}} = \frac{\dfrac{C'A'}{C'B'}}{\dfrac{D'A'}{D'B'}} .$$

We have found the hidden invariance in projecting an arbitrary four points on one line to another perspective with it. What's more, this invariance carries through however long a chain of perspectives we wish to make: if $\ell \overset{o}{\barwedge} m \overset{p}{\barwedge} n$, then,

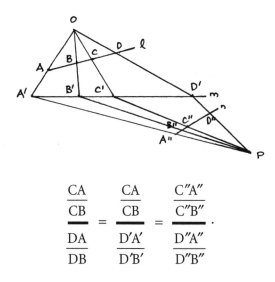

$$\frac{\dfrac{CA}{CB}}{\dfrac{DA}{DB}} = \frac{\dfrac{CA}{CB}}{\dfrac{D'A'}{D'B'}} = \frac{\dfrac{C''A''}{C''B''}}{\dfrac{D''A''}{D''B''}}.$$

Not ratios, then, as with similar triangles, but a *ratio of ratios* is what remains constant in these projective transformations *within* Euclidean geometry. It suggests that the cross ratio is a native of the projective plane, glimpsed here in its travels abroad. The hidden connection it makes (as Heraclitus told us) is stronger than one we can see, but might stand out clearly in its natural setting. To bring it out there, a wonderfully clever way arose of importing coordinates to the projective plane—but to fetch from that far away would need another voyage.

To Chapter Nine

1. [to page 245] The problem with terminating decimals.

Dedekind pointed out a problem with the 1–1 correspondence that Cantor had proposed between the points of the square and those on the line. It involved the awkward fact that "terminating decimals", such as $0.3\bar{0}$ (for $\frac{3}{10}$) can always be represented in another way—in this case, $0.3\bar{0} = 0.29\bar{9}$.

Why is this so? Let's look at a straightforward instance: the number 1. Written in full, this is $1.\bar{0}$. We claim, however—far-fetched as it may seem—that

$$1 = 0.\bar{9}$$

A delightful proof follows just from asking what $0.\bar{9}$ *is*—or to put it mathematically: solve for x when

$$0.\bar{9} = x.$$

Multiply each side of this equation by 10. Since multiplying by 10 shifts the digits one unit to the left,

$$10 \cdot 0.\bar{9} = 9.\bar{9}$$

we therefore have

$$9.\bar{9} = 10x.$$

Now subtract the first equation from the latest one:

$$
\begin{array}{rcl}
9.\bar{9} & = & 10x \\
- \, 0.\bar{9} & = & x \\
\hline
9 & = & 9x
\end{array}
$$

and dividing both sides by 9 reveals that x = 1.

This ambiguity in the naming of a point on the line could be simply resolved by always choosing the "non-terminating" form (in our example, $0.2\bar{9}$). But this choice led to a subtle dilemma. What if the point on the line which we want to match with a point in the square is, say, $.2304\bar{07}$? By Cantor's proposed correspondence this would go to ($0.2\bar{0}$, $0.34\bar{7}$). But since $0.2\bar{0} = 0.1\bar{9}$, written so:

$$(0.1\bar{9}, 0.34\bar{7}),$$

that point must in its turn be sent to the point on the line

$$0.139\overline{497}$$

a very different point from the one originally sent to the point on the plane, destroying the 1–1 correspondence. Since $\leftrightarrow$ runs in both directions, you must return to the point you started out from. But here you don't; you end up quite far away:

$$0.2304\overline{07} \leftrightarrow (0.2\bar{0},\ 0.34\bar{7}) = (0.1\bar{9},\ 0.34\bar{7}) \leftrightarrow 0.134\overline{97}.$$

It looked as if in all his exertions, Cantor had momentarily but disastrously fallen asleep, and as Maurolico commented centuries before: "A little sleepiness and old errors are propagated, new ones introduced." But Cantor awoke with an evasion to equal the snare his friend had discovered. Instead of making the x- and y-coordinates of the point on the plane from the odd and even entries, respectively, of the decimal expression for the point on the line, he preprocessed that expression into *strings*: any non-zero digit is a string, all by itself, and when a zero first appears it begins a new string which ends as soon as a non-zero digit appears. Then send alternate *strings* to these two coordinates.

If the point on the line is, for example, $0.230\overline{407}$, it breaks up into strings like this:

$$0. \; 2 \; \| \; 3 \; \| \; 04 \; \| \; 07 \; \| \; 07 \; \| \; 07 \; \| \ldots$$

and it would be matched to the point in the square with coordinates

$$(0.20\overline{407}, 0.3\overline{07}) \, .$$

Try your hand at finding the coordinates of the point that corresponds to $0.3040005678095\ldots$
Solution: breaking the decimal into strings gives us

$$0.3 \; \| \; 04 \; \| \; 0005 \; \| \; 6 \; \| \; 7 \; \| \; 8 \; \| \; 09 \; \| \; 5 \ldots$$

and this would be sent to the pair of coordinates

$$x = 0.30005709\ldots$$

$$y = 0.04685\ldots$$

2. [to page 256] The set of all subsets of $\mathbb{N}$ has the same cardinality as the set of real numbers.

In order to show that these two sets have the same cardinality, we need to make a 1–1 correspondence between their members.

One of the artful dodges you learn in the guild is to keep simplifying until you actually have to do some work. We can certainly simplify our problem here by recalling (from page 242) that there is a 1–1 correspondence between $\mathbb{R}$ and the elements of the open interval $(0, 1)$—so we need only try to match up the elements of $(0, 1)$ and those of $\mathcal{P}\mathbb{N}$.

The theorem on page 255, proved by Schroeder and Bernstein in 1898, allows us to simplify further. They proved that two sets, A and B, have the same cardinality if and only if

$$\text{card } A \leq \text{card } B$$

and

$$\text{card B} \le \text{card A} .$$

Taking A as $(0, 1)$, B as $\mathscr{P}\mathbb{N}$, the set of all subsets of $\mathbb{N}$, this means that we need only find some way of making a correspondence between each decimal in $(0, 1)$ and a different subset of $\mathbb{N}$; then making a correspondence between each subset of $\mathbb{N}$ and a distinct decimal in $(0, 1)$.

We can no longer put off the actual work, but as often happens, find it attractive once begun. You probably know that any number can be written with just 0s and 1s (this is what your computer does, for which—or whom—0 is a switch in the "off" and 1 in the "on" position). So 0 is 0 and 1 is 1, but 2 is 10, 3 is 11, 4 is 100, because each number is made up by adding together no or one of the successive powers of 2 (17, for example, is $16 + 1$, i. e., $2^4 + 2^0$, or $1 \cdot 2^4 + 0 \cdot 2^3 + 0 \cdot 2^2 + 0 \cdot 2^1 + 1 \cdot 2^0$, so 10001). The Rosetta Stone begins like this:

Numbers base 10	Numbers in Binary
0	0
1	1
2	10 (i.e., $1 \cdot 2^1 + 0 \cdot 2^0$)
3	11
4	100
5	101
6	110
7	111
8	1000
9	1001
10	1010

You can even use this binary notation for decimals:

Fraction	Base 10	Binary	
$\dfrac{1}{4}$	$.25\bar{0}$	$.01\bar{0}$	$\left(= \dfrac{0}{2} + \dfrac{1}{4} + \dfrac{0}{8} + \dfrac{0}{16} + \ldots \right)$
$\dfrac{1}{3}$	$\bar{.3}$	$\overline{.01}$	$\left(= \dfrac{0}{2} + \dfrac{1}{4} + \dfrac{0}{8} + \dfrac{1}{16} + \ldots \right)$

It may take some work to re-express any decimal in terms of fractions whose denominators are successive powers of 2, but it can always be done. We'll need this.

Finally, we need to think of any subset S of $\mathbb{N}$ from a rather interesting point of view. S has various numbers in it: perhaps none, if $S = \varnothing$;

perhaps all, if S = $\mathbb{N}$—but empty, finite, or infinite, we can always say for any natural number n whether n is in S or not. If n *is* in S, we'll assign it the number 1; if it *isn't*, it must bear the mark 0.

This means that to any subset S of $\mathbb{N}$ corresponds an infinite sequence of 0s and 1s. Take the set S = {3, 5, 10}, for example. Here only 3, 5, and 10 have the number 1 assigned to them—all the rest of the naturals get 0. This means that to S = {3, 5, 10} corresponds the infinite sequence

$$0, 0, 1, 0, 1, 0, 0, 0, 0, 1, 0, 0, 0, \ldots$$

We're now ready for the first of our two steps: to show that card $(0, 1)$ ≤ card $\mathcal{P}\mathbb{N}$. We'll do this by sending *each* decimal in $(0, 1)$ to a different subset S of $\mathbb{N}$, as follows. First rewrite the decimal in binary form; then look at that binary form as a sequence of 0s and 1s, and match this sequence to the unique subset S of $\mathbb{N}$ in the way described. So $\frac{1}{3}$, for example, is .010101... in binary: the sequence, that is,

$$0, 1, 0, 1, 0, 1, \ldots$$

which corresponds to the set {2, 4, 6, 8, ...}. In other words, the element $\frac{1}{3}$ in $(0, 1)$ corresponds to the set of even numbers. $\frac{1}{4}$, you'll find, is matched to the subset {2}. In this way, every decimal in $(0, 1)$ corresponds to a different subset of $\mathbb{N}$, so

$$\text{card } \mathbb{R} = \text{card } (0, 1) \leq \text{card } \mathcal{P}\mathbb{N} .$$

It only remains (thanks to Schroeder and Bernstein) to show that card $\mathcal{P}\mathbb{N}$ ≤ card $(0, 1)$. But we really already know how to do this: the subset {3, 5, 10} corresponds, as we saw above, to the infinite sequence

$$0, 0, 1, 0, 1, 0, 0, 0, 0, 1, 0, 0, 0, \ldots$$

so it is the most natural thing in the world (once you have accustomed yourself to this world) to match that sequence to the *ordinary* decimal

$$.0010100001\overline{0}$$

In this way, each and every subset of $\mathbb{N}$ corresponds to a different decimal made up exclusively of 0s and 1s—that is, a decimal between 0.0 and 0.1, which is contained in the interval from 0 to 1—hence, we have card $\mathcal{P}\mathbb{N}$ ≤ card $(0, 1)$. These two parts establish that

$$\text{card } \mathcal{P}\mathbb{N} = \text{card } (0, 1) = \text{card } \mathbb{R} :$$

there are exactly as many real numbers as there are subsets of the naturals.

Appendix

Two slight technicalities may be sticking in your throat, which we will remove by passes more canny than the Heimlich Maneuver. We had been dealing throughout with the *open* interval (0, 1): the set of all decimals between 0 and 1 but not including either 0 or 1. In our last series of steps, however, the empty set would correspond to the sequence

$$0, 0, 0, \ldots$$

and hence to the decimal $.\bar{0}$: i.e., 0, which isn't in (0, 1).

We will simply show (using Schroeder-Bernstein again) that adding this extra element changes nothing, as far as the cardinality goes:

$$\text{card } [0, 1) = \text{card } (0, 1)$$

(where "[0, 1)" means the set of all decimals from 0 to 1, including 0 but excluding 1). We therefore need to show:

$$\text{card } (0, 1) \leq \text{card } [0, 1)$$

and

$$\text{card } [0, 1) \leq \text{card } (0, 1) .$$

The first inequality obviously holds: every element in (0, 1) is also in [0, 1).

As for the second,

$$\text{card } [0, 1) \leq \text{card } \mathbb{R} = \text{card } (0, 1) .$$

Since

$$\text{card } (0, 1) \leq \text{card } [0, 1) \leq \text{card } (0, 1),$$

by Schroeder-Bernstein the two sets have the same cardinality, which is also card $\mathbb{R}$. Hence

$$\text{card } \mathscr{P}\mathbb{N} = \text{card } \mathbb{R} .$$

The other technicality concerns the terminating and the non-terminating forms of binary decimals: the same sort of problem we found and solved in the first part of this chapter's appendix.

The point $0.1\bar{0}$ in (0, 1) can also be written $0.0\bar{1}$. What subset in $\mathscr{P}\mathbb{N}$ do we match it up with? Let's just convene (as we did before) always to

312

choose the "non-terminating" form—so here, $0.0\bar{1}$; and this is unambiguously matched with the subset of $\mathbb{N}$ containing every natural number except 1. Thus indeed

$$\text{card}\,(0, 1) = \text{card}\,\mathscr{P}\mathbb{N}\ .$$

3. [to page 259] Cantor's "proof" that every cardinal number is an aleph.

Having distinguished "consistent collections", or sets, from collections that were in some sense "too large" to be sets—the *inconsistent collections*—Cantor thought he could use the latter to prove what had until then eluded him: that his alephs were the only kinds of cardinals there were. This would mean that the cardinality of power sets too must be alephs, so that they sat ranged in ordered tiers with other sets; and the continuum in particular, which corresponded to the set of all subsets of the natural numbers, would have an aleph for its size—though which aleph this was (perhaps the cherished $\aleph_1$, as he hoped) was still elusive.

To carry through his proof Cantor had to make three assumptions.

Assumption One: Only sets (that is, only consistent collections) have cardinal numbers.

As assumptions go, this one seems plausible enough. It bars from consideration such monsters as the cardinality of the "set" of all sets. You may feel that, in the absence of intuition, this kind of assumption has a merely legal air to it, like a stipulation that asks all parties to the discussion for agreement, just so as to move on to matters more important.

Assumption Two: Two collections with a 1–1 correspondence between their members are either both sets or both inconsistent collections.

If we are going to introduce a distinction into the kinds of collection and still keep the notion of 1–1 correspondence intact, this assumption seems both reasonable and necessary.

Assumption Three: If a collection V has no aleph as its cardinal number, then the whole of Ω (the collection of all the ordinals) corresponds to some subcollection V' of V.

Appendix

When Cantor wrote to Dedekind on July 28, 1899, detailing his proof, he introduced this assumption with the words "We readily see . . ." Such a phrase is notorious among mathematicians as are its companions in infamy, "It is obvious that . . ." and "Now clearly . . ."; they mean that the reader has hours or days of head-splitting labor ahead to bring light to this darkness—after which he may learn that the writer himself no longer remembers why it was obvious.

In addition to these assumptions, Cantor drew on Burali-Forti's result that the collection Ω was inconsistent, and on the reasonable observation (which we have been taking for granted all along) that every subcollection of a set is again a set (in fact, a "subset"). Hence if a subcollection *isn't* a set, neither can be that collection of which it is a part. This means that if a collection X contains a subcollection which is in 1–1 correspondence with Ω, then by Assumption Two this subcollection of X is inconsistent—and hence so is X.

Here then is Cantor's brief proof of the

"Theorem": Every cardinal number is an aleph.

Proof:

1. If a collection V has no aleph as its cardinal number, then (by Assumption Three) the whole of Ω is in 1–1 correspondence with some subcollection V′ of V.
2. Hence (by what we just remarked), V is inconsistent.
3. The contrapositive is: if V is a set (a consistent collection), then V has an aleph as its cardinal number.
4. Hence (by Assumption One), all cardinal numbers are alephs.

In mountains, as Nietzsche pointed out, the shortest way is from peak to peak—but for that you need long legs. Whose are adequate for this crevasse-ridden landscape? Cantor's assumptions in his valiant effort have been thought through and modified since. His "inconsistent collections" are now "classes": collections too big to be sets. Does this make them more thinkable? As you saw, Zermelo's Axiom of Choice is an attempt to bridge all at once over Cantor's third assumption. Would you buy it? A number of axioms have been marshalled from which to derive modern versions of Cantor's proof. How are we to hold their unintuitive truths?

A work of art needn't be finished to be great: "Tell me if ever anything was done!" We are all part of this one's onward expansion.

Bibliography

Bell, E. T. *Men of Mathematics*. Simon & Schuster, 1937.

Cajori, Florian. *A History of Mathematical Notations*. Open Court, 1928.

Coxeter, H.S.M. *Introduction to Geometry*. Wylie, 1969.

Dauben, Joseph. *Georg Cantor: His Mathematics and Philosophy of the Infinite*. Harvard University Press, 1979.

Dunnington, G. Waldo. *Carl Friedrich Gauss, Titan of Science*. Hafner, 1955.

Euclid. *The Elements* (trans. and ed. Sir Thomas Heath). 3rd edition, revised, with additions. Dover, 1956.

Fauvel, John, and Jeremy Gray. *The History of Mathematics: a Reader*. Macmillan, 1987.

Goldman, Jay R. *The Queen of Mathematics: A Historically Motivated Guide to Number Theory*. A. K. Peters, 1998.

Hallett, Michael. *Cantorian Set Theory and Limitation of Size*. Oxford University Press, 1984.

Hardy, G. H., and E. M. Wright. *An Introduction to the Theory of Numbers*. Oxford University Press, 1979.

Heath, Thomas. *A History of Greek Mathematics*. Oxford University Press, 1921.

Katz, Victor J. *A History of Mathematics: an Introduction*. Harper Collins, 1992.

Kirk, G. S., J. E. Raven, and M. Schofield. *The Presocratic Philosophers*. 2nd ed. Cambridge University Press, 1983.

Kline, Morris. *Mathematical Thought from Ancient to Modern Times*. Oxford University Press, 1972.

Meschkowski, Herbert. *Georg Cantor: Leben Werk und Wirkung*. Bibliographisches Institut Mannheim/Wien/Zürich, 1983.

Noether, W., and R. J. Cavaillès (eds.). *Briefwechsel Cantor-Dedekind*. Hermann & Cie., 1937.

Novy, Lubos. *Origins of Modern Algebra*. Noordhoff International Publishing, 1973.

Ore, Oystein. *Number Theory and Its History*. McGraw-Hill, 1948.

Phillips, Esther R. (ed.). *Studies in the History of Mathematics*. Mathematical Association of America, 1987.

Reid, Constance. *Hilbert*. Allen & Unwin/Springer, 1970.

Smith, David Eugene. *History of Mathematics*. Ginn and Company, 1923.

Stillwell, John. *Mathematics and Its History*. Springer-Verlag, 1994.

Van Stigt, Walter P. *Brouwer's Intuitionism*. North-Holland, 1990.

Zagier, Don (trans. R. Perlis). "The First 50 Million Prime Numbers." *Mathematical Intelligencer*, v. 0 (1977): 7–19. The original is "Die Ersten 50 Millionen Primzahlen," in *Mathematische Miniaturen* (Birkhaeuser, 1998): 39–73.

Non-Print Media

www-history.mcs.st-and.ac.uk

Index

Note:
— Mathematical symbols can be found under their name (i.e., infinity). They can also be found collectively under the "symbols" entry.
— Page numbers in **bold** refer to entire sections.
— Page numbers in *italics* refer to illustration captions.

a priori knowledge, 33, 52–53
Abel, Niels, 96
absolutely infinite, 259
abstraction, 13, 132, 242
Abu 'Ali al-Hasan ibn al-Haytham, 46
addition: additive identity, 36, 266; additive inverse, 36, 40, 58–59, 266, 267; alephs, 257; axioms of, 59; closure, 37; commutative axiom, 96–97; complex plane, 178; field axioms, 38, 264–65; imaginary numbers, 26–27; infinite numbers and, 97; integers, 14; lengths, 147; natural numbers and, 25; number "1" and, 5; number shapes and, 11; odd numbers, 45; of plane coordinates, 243; polygon construction and, 147; square roots of negative numbers, 169; sums, 77–78, 89, 92, 263–64
aesthetics in mathematics, 133
A'h-Mose (scribe), 126
Alberti, Leon Battista, 202–4, 205, 217
Alberti's Veil, 202–4, 213–14
Alcibiades Humor: alternate perspectives, 116; arrogance of, 130; Cantor, 229; as characteristic of mathematicians, 47, 202; completing the square, 286; described, 25–26; imaginary numbers and, 173, 192; interpretation of straight lines, 125, 127; intuition and, 53; play as characteristic of, 89
Alembert, Jean le Rond d', 173, *173*
alephs ($\aleph$). *See also* transfinite numbers: Cantor and, 246–47, 254; cardinality, 256–57, 259
algebra: algebraic form of lines, 155; complex numbers and, 171–72; complex plane and, 173; conceptions of infinity, 132; coordinate plane, 154; cubic roots, 292–300; eagle of algebra, **133–66**; formulation of circles, 159–60; Fundamental Theorem of Algebra, 169, 173, 193–94; geometry and, 174, 177–78;

Golden Ratio, 137–38; pentagon construction and, 140; projective geometry and, 227; quadratic equation, 162, 285–86, 286–88; square root extension fields, 157; symmetry in, 286
algorithms, 23–24
ampliatio, 42
Anaximander, 75
"Ancient Days of Old" (Blake), 133
angle-bisectors, 111, 208–9
angle-side-angle (ASA) postulate, 104–5, 109
animals, 3
Annex, on-line, 66, 85
apeiron, 75, 258
Aquinas, Thomas, 32
Archimedes, 82, 114, 125, 133, 188
area, 71, 90, 114
Argand, Jean Robert, 177, 180
Argand diagrams, 177
Aristotle, 31, 100, 236
Arithmetica (Maurolico), 45
Armstrong, Louis, 163
arts: mathematical, 2; music, 2, 16, 138; painting, 202; perspective drawing, 7, 202–4, 215–16
Ashford, Daisy, 248–49
associative axiom, 56–57, 97
associative law, 34, 35, 38
asymmetry, 11, 204, 205
Atiyah, Michael, 132
Axiom of Choice, 256, 314
axioms, **29–55**; "1" and "0," 35; addition of infinite numbers, 97; additive identity, 36; additive inverses, 36, 37, 40, 58–59, 266, 267; axiomatic systems, 266; closure, 37; consistency, 50; discovery *vs.* invention of, 33, 44; for fields, 38, 49; as foundation of mathematics, 31; Gergonne on, 32; Hilbert's, 54; multiplicative identity, 36; multiplicative

axioms (*continued*)
  inverses, 36, 37, 56–57; multiplying by zero, 40; Peano's Axioms, 46, 54, 264–65; projective geometry, 206–7, 209, 211, 212; questions arising from, 56; truth of, 52
axis of perspectivity, 219

Babylonians, 90
Bacon, Francis, 9, 11, 257, 261
Bhāskara, 287
balance, 112–13
Baudelaire, Charles Pierre, 1, 131
Blake, William, 133, 199, 262
Bombelli, Rafael, 170, 180, 256
"The Book," 61
Brahe, Tycho, 44–45
Breughel, Pieter, 138
Brouwer, L. E. J., *48*; alephs and, 257; "empty form," 44; Hilbert and, 52–55; mathematical intuition, 47–49, 271; on sequences, 88–89
Brown, J. B., 58, 59, 170
Burali-Forti, Cesare, 258, 314
Burali-Forti paradox, 258–59, 314

Cairn of Remembrance, 232
calculus, 71, 125, 188, 189
Calgacus, 167
Cantor, Georg, *240, 261*; alephs, 254–55, 313–14; background, 229; Burali-Forti paradox, 259; completed infinities, 236; Continuum Hypothesis, 261, 262; counting, 241; decimal numbers, 239; Dedekind and, 243–45, 314; Hilbert's defense of, 258; set theory, 232, 248–49; terminating decimals, 308–9; use of correspondence, 236–37; use of diagonals, 235
Cardano, Girolamo, 170, 241, 294–95, *295*
cardinality: 1-1 correspondence, 231–33; alephs, 256–57, 313–14; Burali-Forti paradox, 259; Continuum Hypothesis, 261; line segments, 240–41; Schroeder-Bernstein theorem, 309–13; set of all subsets, 309–13; of sets, 248, 259; of squares, 245–46; of varieties of numbers, 250–56
Cartesians, 32
Cauchy, Augustin Louis, 192
center of gravity, 112
center of perspectivity, 214, 220
central angles, 136, 138–39
centroids, 112, 113–14, 119, 277–78
chaos, 66, 67, 75–76, 167
choice, 256, 314
circles: algebraic formulation of, 159–60; chords, 264; circumcircles and circumcenters, 109; complex numbers and, 196–97; Euler Circle, 123; Feuerbach Circle, 123; incenter and incircle, 111; incircle tangent, 208–9; nine point circle, 123–24, 278; non-Euclidean geometry and, 264; polygon construction and, 145–47, 150–51; in projective geometry, 225; radians, 189–90; sine and cosine, 193; square root extension fields, 154, 158–61; triangles and, 108, 135–36, 142
circumcenters, 109, 118, 119, 276–77
Clark, Kenneth, 131
Clarkson, James, 270–76

clocks, 51
closure, 37, 38, 173
Cohen, Paul, 262
collections, 39, 229, 259, 314
collinearity: collinear lines, 223, 225; collinear points, 120; in projective geometry, 216
common chord, 163
commutative axiom, 96–97, 267
commutative law, 33, 35, 38, 58
complementary congruence, 109–10
completing the square, 285–86, 286–88
complex numbers: addition of, 178; conjugates, 180; cubic roots and, 170–72; de Moivre on, 194–95; Dedekind Cuts, 39–40, 44, 49, 179; geometrical representations, 188; in hierarchy of numbers, 28, *28*; "i" (square root of –1), 169–70, 171–73; modulus, 179, 180, 183; multiplication in a+bi form, 178–80; multiplication in r (cos φ + i sin φ) form, 184–88; Pythagorean Theorem and, 42; trigonometry and, 189, 296
complex plane, *167–99*; closure under algebraic operations, 173; complex numbers and, 174–76; cube roots and, 193–94; de Moivre's Formula, 198; illustrated, 27; multiplication on, 178–79; product vector, 180; sine and cosine, 183
complex roots, 169, 170, 173, 194, 199, 290–91
composite numbers, 59
computers, 51
congruence, 215
conic sections, 226
conjugates, 180
consistency, 50, 264–66
constants, 86
Continuum Hypothesis, 256, 257, 261
convergence: convergent series, 94–98; infinite series, 192; limit points, 90; prime reciprocals, 271–73; square root of 2, 99
coordinate systems: coordinate plane, 153–54, 154–55; projective geometry and, 227; terminating decimals and, 309
Copernicus, 181
coplanar triangles, 218–19
correspondence (1–1), 231–44, 313–14
cosine (cos), 182–88, 190–97, 296, 301
Cotes, Roger, 194
countability, 232–34, 242, 253–54
counting numbers: cardinality, 250–51, 253, 255–56; correspondence, 234–35, 238; counting primes, 73; nursery rhymes, 3–4; positional notation, 7; primes, 62; sets, 232; zero, 8
creation, 8. *See also* invention
cubic curve, 41
cubic functions, 168
cubic roots: complex numbers and, 170–72; complex plane and, 193–94; de Moivre and, 197; fields and, 152; polynomials and, 292–300; rational numbers and, 288
cupping technique, 245–46
curiosity, 200

Dauben, Joseph Warren, 261
de Moivre, Abraham, 194–95, 197, 198

de Moivre's Formula, 198
debt, 12–13
decagon, 138, 139–40, 143
decimals, 21–22, 238–39, 307–9
Dedekind, Richard: Cantor and, 237, 243–45, 261, 314; Formalism and, 37; mathematical induction, 46; on sets, 39–40; terminating decimals, 307–9
Dedekind Cuts, 39–40, 44, 49, 179
Democritus, 42
"Desarguean configurations," 220
Desargues, Girard, 215, 216–17, 218–19
Desargues's theorem, 220–21, 226
Descartes, Rene: complex plane, 27; coordinate plane, 153; factor theorem, 292–94; form of circles, 158; formalism, 38; mathematical intuition, 32; proof methods, 92
diagonals: correspondence, 237, 239; Gödel's use of, 262; "legitimate construction," 204; sets theory and, 249
diagrams, 2
Diarium (Hermes), 288–90
Dirichlet, Johann Peter Gustav, 67
discovery *vs.* invention, 8, 25, 33, 47–48
distance, 125
distributive axiom, 38, 40, 58–59, 267
distributive law, 35, 38, 44
divergent series, 94–98, 270–76
division: decimal representation and, 21–23; imaginary numbers, 26–27; of plane coordinates, 243; polygon construction and, 148; rational numbers and, 25; square root extension fields, 156; by zero, 56–57, 57–58
dodecagon, 144
drawing, 2, 7, 202–4, 215–16
Du Bois-Reymond, Paul, 245
Duke of Wellington, 157

eagle of algebra, **133–66**
economics, 16
Egyptians, 16, 90
Einstein, Albert, 227, 245
Elements (Euclid), 104
elements of sets, 247–48, 249
ellipses, 225
"empty form," 44
empty sets, 248
endpoints, 241
Enlightenment, 34, 119
Eötvös, Lóránd, 128
equality: of complex numbers, 171; in projective geometry, 215
equilateral triangles, 112, 123–24, 135, 197, 283–84
Erastosthenes, 61–62
Erdös, Paul, *61*, 98
Euclid and Euclidean geometry, **100–130.** *See also* projective geometry; calculus and, 188; congruent triangles, 104–5; consistency, 264; construction of polygons, 133–35; Euclidean plane, 104–5, 166, 202; Euclidean representation, 209; Euclid's Fifth Postulate, 100–104; Fermat numbers and, 199; infinite primes, 60–61; Millay on, 124; multiplication of negatives, 267–68; paradoxes of, 49; parallel

lines in, 100–104, 114; perspective and, 204; polygon construction, 147; prime numbers and, 65; prime reciprocals, 273–74; shortest path in triangles, 282; straight lines in, 125; sum of ratio series, 90; triangle's role in, 106–7, 117, 144
Eudoxus, 31
Euler, Leonhard, *119*; on complex numbers, 174; complex variables, 192; Euler Circle, 123; Euler Line, 119, 123–24, 278; exponential functions, 269; infinite series proof, 98; mathematical induction, 47; p of geometry, 71; reciprocals of primes, 270–76
"Ex Oriente Lux," 262
exponential functions, 70, 98, 191–92, 268–70

The Factor Theorem, 292–94
factorial, 65, 191
factoring, 286
factors, 60, 268, 292–93
Fagnano's Problem, 124–30, 284
false position, 126
Farquharsons, 232
Fejér, Leopold, 127–28
Fermat, Pierre de: coordinate plane, 153; Fermat numbers, 165, 199; Fermat point, 279–85; Fermat prime, 288; form of circle, 158
Feuerbach, Karl Wilhelm, 121, 124
Feuerbach Circle, 123
fields. *See also* square root extension fields: axioms of, 38, 49; complex numbers and, 171; coordinate plane, 154; cubic roots and, 152; field axioms, 264–65; Formalism and, 37; polygon construction and, 150–51; slope in, 156–57; square root extension fields, 163; Weber's tablets for Fields, 149–50; y=mx+k form of line, 155–56
Formalism: Axiom of Choice and, 256; axiomatic systems, 29–55; Cantor and, 258; consequences of, 265; fields and, 37; intuition and, 130; natural numbers, 42; non-Euclidean geometry and, 35
fourth roots, 151–52
fractions, 15–17, 23, 233, 234, 236
Freemasonry, 261
Frege, Gottlob, 51
functions, 53, 70, 189
Fundamental Theorem of Algebra, 169, 173, 193–94

Galileo Galili, 228, 229, 233, 240
gambling, 86
Gaudi i Cornet, Antoni, 133
Gauss, Karl Friedrich, *68*; 65,537-gon, 288; approximation of π (x), 71–72; completed infinities, 236; complex variables, 192; Fundamental Theorem of Algebra, 169; influence, 163–64; mathematical intuition, 31; prime numbers and, 67–68, 69–71; square root extension fields, 164; square root of –1, 188; triangular numbers, 82; varieties of geometries, 52
geometry, **100–130.** *See also* projective geometry; *specific shapes*; algebra and, 174, 177–78; geometric sequences and series, 86, 89, 96, 132;

Index

geometry (*continued*)
geometrical representations, 188; Greeks and, 18; Hobbes on, 104; paradoxes of, 49; parallel lines, 52; perspective and, 204; projective, 302–7; Pythagorean Theorem, 29–30; straight lines in, 125; symbols and abbreviations, 105–6; variety of, 52, 133

Gergonne, Joseph Diez, 32

Gerson, Levi ben, 45

Gilbert, Humphrey, 173

Girard, Albert, 27–28

Gnostics, 32, 48

Gödel, Kurt, 54, 262, 265

Gödel's Incompleteness Theorem, 54

Goethe, Johann Wolfgang von, 184

Goldbach, Christian, 73

Golden Mean and Ratio, 137–38

Goya, Francisco Jose de, 96

graphs, 68–69, 182

Graves, Robert, 100

gravity, 36

Great Chain of Being, 151

"The Great Converse," 51

Great Unconformity, 262

Greeks: *apeiron*, 75; geometry, 18, 101, 133; insight and intuition, 31; number theory and, 8; ratios, 14–16

growth functions, 70, 98, 191–92, 268–70

Hadamard, Jacques, 71

Haldane, J. B. S., 221

Halmos, Paul, 178

Hamilton, William Rowan, 39, 53, 167–68

Hamlet (Shakespeare), 205

Harmonic Series, 94–96, 270–76

Hawking, Stephen, 2

al-Haytham, Ibn, 46

heptagons: constructing, 134, 166, 198–99, 291–92, 292–300, 300–302; heptagonal numbers, 85

Heraclitus, 9, 120

Hermes, Johann, 165, 288–90

hexagons: constructing, 143–44, 198; hexagonal numbers, 82

hierarchy of numbers, *28*

Hilbert, David, *49*; alephs and, 255; axiomatic systems, 49–55; axioms, 147; Cantor and, 240, 258; infinite series proof, 98; on mathematical reasoning, 178; on Pappus's theorem, 226; on Peano's Axioms, 265; on philosophy of mathematics, 35; projective geometry, 206; rivalry with Brouwer, 47

Hipparchus, 181

Hippasus, 18–21, 25, 31, 138, 140–41, 262

Hobbes, Thomas, 85, 93, 103–4, 124, 133

Hofmann, J. E., 280–81, 282

horizontals, 205

Hutton, James, 262

hyperbolas, 226

hypotenuse, 21, 108, 110–11, 182, 276–77

"i" (square root of –1), 171–73. *See also* imaginary numbers

identity, 38, 59

images of numbers, 4, 27–28

imaginary numbers: Bombelli and, 256; complex plane and, 174; complex roots of 1 and, 290–91; exponential functions and, 192; "i" (square root of –1), 171–73; mathematical operations on, 26–27

imagination, 173–74, 246

incenter, 111, 208–9

incircle, 111, 208–9

Incompleteness theorem (Gödel), 54

"indefinite," 75

Indian culture, 191

induction, 42–46, 51, 263–64, 265

infinity: absolutely infinite, 259; cardinality of, 237, 251; collections of numbers, 229; completed infinities, 236, 252; conceptions of, 1; counting numbers, 6; Dedekind Cuts and, 49; exponential growth, 191–92; infinite area, 92–93; infinite sequences and series, 88, 92, 93–95, 116, 189, 192; infinite sets, 248; "more than infinity," 236; natural logarithms, 95; natural numbers, 230–31; perspective and, 204; potential infinities, 236; prime numbers, 60; projective geometry and, 223–25; proof of, 5–6; ratios and, 17

inscribed plane, 208–9

inspiration, 229, 254–55, 257–58

integers: 1–1 correspondence, 232; cardinality, 232–33; countability, 242; in hierarchy of numbers, *28*; paired, 41; polygon construction and, 149; as rationals, 17

integrals, 71

interleaving, 244

intersections, 210, 218

Introductio Arithmetica (Nichomachus of Gerasa), 13

intuition. *See also* Alcibiades Humor: alephs and, 257–58; Brouwer, 47–49, 271; Descartes, 32; formalism and, 34, 50–53, 130, 147; Gauss, 47; imaginary numbers and, 188

invention, 25, 33, 47–48

Invercauld, 232

inverses: additive, 36, 40; field axioms, 38; multiplicative, 36

irrational numbers: abstract nature of, 38–39; correspondence, 237, 238; cubic roots and, 297; cyclic, 23; decimal form, 21–23; Dedekind Cuts and, 39–40, 49; discovery *vs.* invention of, 25; exponential functions, 70, 269; functions and, 53; logarithms, 70, 95; polygon construction and, 149–51, 151–53; square root of 2, 20–21

Jerusalem, 131

Jia Xian, 248

John of Austria, 44

Joseph of Arimathea, 262

Joyce, James, 167

Jung, Carl, 31

k-gons. *See* polygons

Kant, Immanuel, 33, 35, 52–53

al-Karaji, Abu Bakr, 45–46, 248

Keller, Wilfrid, 165

Kepler, Johannes, 138

Keyser, C. J., 213

Index

König, Jules, 257
Kronecker, Leopold, 13–14, 42, 67, 236, 255

Lambert, Johann Heinrich, 205
Langland, William, 149
language of mathematics, 4, 47, 56
Leibniz, Gottfried Wilhelm, 27
lemma, 278–79
length in complex plane, 179, 180
Leonardo da Vinci, 76
"less than minus," 170
limits, 87–90, 96–97, 252, 258, 273
lines: algebraic form, 155–56; correspondence
    with points, 244–45; line segments, 240–41;
    linear thinking, 235; parallel, 100–101; in
    projective geometry, 208, 210–13, 218, 219,
    221–23, 223–24, 302–7; slope, 156–57, 162
Linnaeus, Carolus, 169
Littlewood, J. E., 91, 289–90
Littlewood's Miscellany (Littlewood), 289–90
Li(x), 71–72
Lloyd, Harold, 235
Lobkowitz, Juan Caramuel, 134
logarithms, 70, 72, 95, 268–70
lucky numbers, 4

magnitudes, 16
Mahavira, 1
major sixth interval, 138
Martingale System, 86
mass, 114, 277–78
"mathematical engine," 34
Mathesis Audax (Lobkowitz), 134
Maurice of Nassau (Maurice of Orange), 32
Maurolico, Francesco, 44–45
mean proportional, 142, 174
median, 111–12
Mendelssohn, Rebecca, 67
Menelaus of Alexandria, 181
metaphysics, 258
Michelangelo, 8
Millay, Edna St. Vincent, 124
Miloradovich, Mikhail, 204
Minkowski, Hermann, 95, *95*, 99
Möbius, Augustus Ferdinand, 304
modulus, 179, 180, 183, 197
moon, 45
"more than infinity," 236
"more than minus," 170
Moses, 131
multiplication: alephs, 257; axioms of, 56–57,
    59–60; closure, 37; complex numbers, 178–
    84; field axioms, 38, 264–65; imaginary
    numbers, 26–27; integers, 14; multiplicative
    identity, 36; multiplicative inverses, 36, 55–
    57, 271; natural numbers, 231; negative
    numbers, 57–58; plane coordinates, 243;
    polygon construction and, 147–48; square
    roots, 38; square roots of negative numbers,
    170; by zero, 7–8, 40
Murray, Gilbert, 100
music, 2, 16, 138

n-dimensional space, 246
n-gons, 144, 198. *See also* polygons

"nameless" numbers. *See* irrational numbers
Napier, John, 70
natural logarithms, 95, 269–70
natural numbers: 1–1 correspondence, 232–33,
    235, 237; associative law, 34; axioms, 49–50;
    cardinality, 251, 253; commutative law, 33;
    complex plane and, 195–96; cube roots and,
    288; functions and, 53; in hierarchy of num-
    bers, *28*; inductive proofs, 42–43; infinity
    and, 228, 242; multiplicative inverses, 271;
    ordinals and, 251–52; Peano's Axioms, 46;
    primes and, 60, 61–62, 64–65; Pythagorean
    Theorem, 18–21, 42; real numbers compared
    to, 240; sequences, 78–79; squares, 229–30;
    sums, 29, 77–78; time and, 53; triangular
    numbers, 8–9
negative numbers: 1–1 correspondence, 232–33,
    235; abstract nature of, 13–14; debt, 12–13;
    multiplying, 25, 57–58, 267
Neumann, John von, 24
Newton, Isaac: calculus, 188; complex plane, 27;
    formalism, 36; infinite series, 191–92; on
    vastness of mathematics, 77; Wordsworth
    on, 49
Nietzsche, Friedrich, 228, 314
Nine Point Circle, 123–24
non-Euclidean space, 210
numbers. *See also specific types of numbers (i.e.,
    natural numbers, prime numbers, etc.)*: ab-
    straction of, 13; cardinal, 231 (*see also* cardi-
    nality); collections of, 229; coordinates,
    154–55; countability, 232–34, 242, 253–54;
    counting numbers, 4, 6, 8; counting primes,
    73; decimal form, 21–24; Dedekind Cuts
    and, 39–40, 44, 49, 179; even numbers, 19–
    20; hierarchy of, *28*; images of numbers, 27–
    28; lucky numbers, 4; names of, 4, 6–7;
    number line, 149; numerical coordinates (*see*
    coordinate systems); one, 5; patterns, 8, 10;
    positional notation, 7; shapes of, 8–10; signs
    of, 36; tailing primes, 73; triangular, 8, 17–
    18; uncountable numbers, 237; unnatural
    numbers, 12; varieties of, *3–28*; zero, 7
nursery rhymes, 3

objectivity, 215
obtuse triangles, 108, 129–30, 213–14, 278–79,
    284–85
occult, 138
odd numbers, 12–13, 45
Odysseus, 162
Ohm, Georg, 33–35, 36, 67
Ohm, Martin, 33–34
Oksapmin of New Guinea, 6–7
omega (Ω), 254, 314
omega (ω), 251–52
on-line Annex, 66, 85
"On the Analytic Representation of Direction;
    An Attempt" (Wessel), 175
On the Psychology of Military Incompetence
    (Dixon), 47
open intervals, 237–42, 312
ordering, 233–34
ordinal numbers, 251–52, 253–54, 256–57
Oresme, Nicole d' (Bishop of Lisieux), 92, 94–95

Index

organic geometry, 204
organic growth, 70, 269
orthocenters of triangles, 118–19, 121

Paine, Thomas, 8, 17
painting, 202
Palmanova, Italy, 166
Pappus of Alexandria, 140, 225, 226, 282, 304
parabolas, 226
paradigms, 200
paradoxes, 49–50, 258
parallel lines: convergence, 132; on coordinate
    plane, 154–55; Euclid's Fifth Postulate, 100–
    104; non-Euclidean geometry, 264; parallel
    postulate, 100–104, 116–17; polygon con-
    struction and, 148; projective geometry and,
    205, 220, 223–24; proofs, 114–15
parallelograms, 114–15, 117–18, 122, 176
partial sums, 86–87
Pascal, Blaise, 34, 225, 248
patterns: Cantor on, 245; intuition and, 10;
    natural numbers, 77–78; prime numbers,
    66–67; ratios, 87–88
Patterson, Paddy, 288
Peacock, George, 34, *34*, 93–94, 149
Peacock's Principle of Permanence, 34–35, 51–
    52, 93–94, 227
Peano, Giuseppe, 46
Peano's Axioms, 46, 54, 264–65
Peirce, Benjamin, 193
pentagonal numbers, 80–82, 85
pentagons, 136–46, 163–64, 165, 198
pentagrams, 137, 138
perpendicular bisectors, 108, 110
perpendicular lines, 106–8
personality in mathematics, 116, 229, 260. *See
    also* Alcibiades Humor
perspective: axis of perspectivity, 219; cardinal-
    ity, 240–41; center of perspectivity, 214, 220;
    drawing, 202–4, 215–16; projective geometry
    and, 214, 217, 219–20, 302–7; vanishing
    point, 7
philosophy of mathematics, 31
physics, 116
pi (π), 68, 71–72, 189–91, 196–97
pictures as proofs, 91–92
Piers Plowman (Langland), 149
planes and plane geometry: cardinality, 246;
    conic sections, 226; coordinate plane, 154;
    Pappus's theorem, 226; projective geometry
    and, 206, 210–11, 215–18, 220, 222–24; se-
    quences and, 100; triangle as fundamental
    unit, 134
Plato, 31, 50
Playfair, John, 262
Plutarch, 26
Poetics (Aristotle), 100
Poincaré, Henri, 51, 96, 124, *124*, 255
points: correspondence with lines, 244–45; in
    projective geometry, 211–13, 219, 302–7
polygons: complex plane, 198–99; constructing,
    133–34, 136–46, 147, 149–53, 164, 165–66,
    288–92; pentagonal numbers, 80; polygonal
    sequence, 83–84; prime numbers and, 165;
    projective geometry and, 205; regular

n-gons, 134, 144–45, 198–99; triangle as
    simplest form, 104–5; as two dimensional
    objects, 200
polyhedra, 200
polynomials, 168, 169, 189, 292–300
polytopes, 200
Poncelet, Jean-Victor, 204, *204*, 213, 227
positio falsa, 126, 128, 137
positional notation, 7
positions, 149
positive rationals, 234–35
postulates. *See also* axioms: angle-side-angle
    (ASA) postulate, 104–5, 109; axiomatic sys-
    tems, 50; Euclid's Fifth Postulate, 100; paral-
    lel postulate, 100–104, 116–17;
    side-angle-side (SAS) postulate, 105, 107
Poussin, Charles Jean Gustave Nicolas de la
    Vallée, 71
power sets, 248, 249–50, 256
prime numbers, **59–74**; chaos and, 67; counting
    primes, 73; distribution of, 62–65, 68–69;
    divergent series, 270–76; factorial (!), 65; fac-
    tors, 292–93; Fermat numbers, 165; Fermat
    prime, 288; infinite series and, 98; polygons
    and, 151, 165; prime reciprocals, 270–76;
    prime sequences, 268; tailing primes, 73;
    twin primes, 66, 72–73
Princess Ida, 161
Principle of Continuity, 227
projective geometry, **202–27**; collinearity and,
    216; correspondence and, 241; Euclid and,
    207; impact of, 227; infinity and, 223; invari-
    ants on, 302–7; "Kooshball" example, 210;
    "pencil," 206; Poncelet and, 205; projective
    plane, 206, 208–10; projective three-space,
    216, 220
proofs: addition of ratio series, 88–90; additive
    identity, 266; associativity, 266; by contradic-
    tion, 20, 56, 237, 249, 250, 288; correspon-
    dence, 245; Euler Line, 119; of heptagon
    construction, 291–92; imagination and, 246;
    inductive, 42–44; infinite cardinalities, 240;
    k-gonal sequences, 85; multiplication of
    negatives, 267–68; pentagonal numbers, 80;
    personality and, 30–31; pictorial representa-
    tions, 91–92; prime numbers and, 61, 67;
    Pythagorean Theorem, 29–30; quadratic
    equation, 285–86, 286–88; reciprocals of
    primes, 270–76; sum of odd numbers, 45;
    sum of ratio series, 90; sum of the first n
    odd integers, 263–64; symbolic proofs, 78–
    79, 105–6; terminating decimals, 307–9; tri-
    angles, 106–7, 112–16, 121–24
Proust, Marcel, 212
Ptolemy, 181
Punch, 58
pyramids, 217–18
Pythagoras, 16, 158, 180, 183, 262
Pythagorean Theorem, 18–21, 29–30, 104, 138,
    140–41, 158–59
Pythagoreans, 18, 19–21, 32

Quadratic Equation, 140–41, 161–62, 285–86
quadratic functions, 168–69
quartic functions, 168–69

radians, 189–90, 195, 197

rational numbers: coordinate plane, 154; correspondence, 235–38; cube roots and, 288; "Dedekind Cuts" and, 39–40; density of, 21; division and, 25; functions and, 53; in hierarchy of numbers, 28; infinity and, 233, 242; integers as, 17; polygon construction and, 148, 149, 151; positive rationals, 233, 234–35; Pythagorean Theorem and, 42; ratios and, 88; square root extension fields, 156; time and, 53

rational roots theorem, 293–94

ratios: infinity and, 17; limits and, 86–89; polygon construction and, 148; Pythagorean Theorem, 18–21; rational numbers and, 14–16; triangular numbers and, 17; twin primes and, 72

Ravel, Maurice Joseph, 132

real numbers: cardinality of, 250–51; compared to naturals, 240; complex numbers and, 170; complex roots of 1 and, 290–91; composition of, 24; coordinate plane, 154; correspondence, 237, 238–39; "Dedekind Cuts," 39–40; in hierarchy of numbers, 28; imaginary numbers, 28; infinity and, 242; modulus, 179; power sets and, 248; Pythagorean Theorem and, 42; real number line, 241–42

rectangles, 137–38

reflection, 126

"Reflections on the General Cause of Winds" (d'Alembert), 173

Règles pour la direction de l'esprit (Descartes), 32

relativity, 245

religion and theology, 258, 261, 288

revolutions, 262

Rhind Papyrus, 126

Richelot, Friedrich Julius, 165, 288

Riemann, Bernard, 47

Rimbaud, Arthur, 32, 79

Roland, Childe, 137

Romanticism, 34

roots, 25, 38, 168–69. See also square root extension fields; squares and square roots; complex, 169, 193–94; cubic roots, 152, 170–72, 193–94, 197, 288, 292–300; fourth roots, 151–52; of "i," 173; multiplication and, 38; rational roots theorem, 293–94; roots of unity, 198

Rosicrucians, 261

Russell, Bertrand, 260

Saccheri, Girolamo, 205

St. Augustine, 32

Saladin (Salah-al-Din Yusuf ibn-Ayyub), 131

al-Samaw'al, 46

Saracens, 131

Schopenhauer, Arthur, 255

Schroeder-Bernstein theorem, 309–13

Schumann, Robert, 34, 37, 304

Schumann problem, 265

Schwendenwein, 165, 288

self-evident truth, 32–33, 49

sequences and series, 77–99; convergent, 94–95, 96–97; defined, 79; divergent, 94–95, 96; on Euclidean plane, 100; geometric, 86–89, 94,

96, 132; Harmonic Series, 94–96, 270–76; infinite, 88, 92, 93–95, 116, 189, 192; k-gonal sequences, 83–84, 85; Peacock on, 93–94; polygonal, 83–84, 85; prime sequences, 268

series. See sequences and series

sets and set theory, 228–62; 1-1 correspondence, 231, 232, 237; alephs, 246–47, 255, 259, 313–14; cardinality, 247–49, 249–51; consistent and inconsistent, 259; contradictions of, 260–61; defining, 46; as fundamental, 39; infinite sets, 248; paradoxes of, 49, 258; power sets, 248; set of all subsets, 309–13

Shandy, Tristam, 257

Shelly, Mary, 164

side-angle-side (SAS) postulate, 105, 107

signs of numbers, 36

simplicity, 167

sine (sin), 181–88, 190–97, 296, 301, 306

slope, 156–57, 162

Socrates, 25–26, 75

solids, 93

Solon, 75

Sophocles, 189

sorcery, 137

space: cardinality, 246; n-dimensional space, 246; non-Euclidean, 210; three dimensional projective space, 216, 220; variety of geometries, 52

Spinoza, Baruch, 200

square root extension fields, 151, 153–59, 159–61, 161–64

squares and square roots. See also square root extension fields: addition, 12; cardinality, 245; circles and, 198; complex numbers and, 171; construction of, 133–34, 135–36; cube roots compared to, 193; imaginary numbers and, 172–73; irrational numbers and, 21; multiplication and, 38; natural numbers and, 228, 229–30; negative numbers and, 25; pentagonal numbers and, 85; polygon construction and, 144–45, 151–52; reciprocals of primes, 99; square numbers, 9–11, 79–81

statistics, 194

Stevenson, Robert Louis, 5

Stoic philosophers, 31, 32, 48

strings of decimals, 309

structure, 247, 264–66

subcollections, 314

subfield, 150

subscripts, 238

subsets, 247–51, 309–13, 314

subtraction: complex plane, 176–77; debt, 12–13; imaginary numbers, 26–27; lengths, 147; natural numbers, 12–13; negative numbers, 25; of plane coordinates, 243; polygon construction and, 147; square root extension fields, 156; square roots of negative numbers, 169

surveying, 175–76, 177

Sylvester, J. J., 210

symbols: alephs (ℵ), 246–47, 254, 256–57, 259; factorial (!), 65; infinity (∞), 90; mathematicians' use of, 2; "n," 10; omega (Ω), 254, 314; omega (ω), 251–52; Peano and, 46; pi (π), 68, 71–72, 189–91, 196–97; polygonal sequences, 85; in proofs, 30–31, 105–6; symbolic proofs, 78–79 symmetry: in algebra, 286; complex plane, 176; functions and logarithms, 270; in number shapes, 11; symbolic proofs, 78–79

Tablet of the Law, 38
tailing primes, 73
tangent lines, 111
Tanton, Jim, 278–79
tautology, 32–33
terminating decimals, 307–9
tetractys, 18, 28
Thabit ibn Qurra, 228, 257
Thales, 120, 124, 142, 282
Thales's Converse, 120, 122–23, 142
theorems, 29–55. *See also specific theorems*
Theory of Everything, 200
Theosophy, 261
three-dimensional space, 216, 246
Thurston, William P., 174
time, 3, 6, 13–14, 53
Torricelli, Evangelista, 93, 281
Tower of Mathematics, 138
transfinite numbers, 252–56, 260
transitive property, 108, 115, 117, 121, 158, 161, 236
trapezoids, 91
triangles: altitudes, 116, 118–19, 121–24, 128–30; base angles, 140; bisectors, 108; centroid, 112–15, 277–78; circle's relationship to, 121–24, 135–36, 142; circumcircle and circumcenter, 109, 276–77; compared to other polygons, 200; congruent, 104–5, 109–10, 144; construction of, 133–34; coplanar, 218–19; cube roots and, 197; equilateral triangles, 112, 123–24, 135, 197, 283–84; Euclidean geometry and, 104–5, 106–7; Euler line, 119; Fermat point, 279–85; hypotenuse, 21, 108, 110–11, 182, 276–77; incenter and incircle, 111; isosceles triangles, 225; Jia Xian's triangle, 248; median, 113–14, 115–16; obtuse triangles, 108, 129–30, 213–14, 278–79, 284–85; orthocenter, 118–19, 121; parallel postulate and, 100–104; pentagon construction and, 139–40; polygon construction and, 142; prime numbers and, 165; projective geom-

etry, 213–16, 218, 225; Pythagorean Theorem, 18–21; ratios and, 15–16; right triangles, 18–21, 105, 108, 110–11, 120, 123–24; shortest paths in, 127–30, 278–79; sine and cosine, 183; transivity of equality, 108; triangular numbers, 8–10, 18, 79–81, 82
trigonometry, 180–81, 189, 194–95, 241, 304–6
truth, 32–33
twin primes, 66, 72–73

uncertainty, 261
uncountable numbers, 237
Une saison en enfer (Rimbaud), 32
unnatural numbers, 12

vanishing point, 7, 203, 205
vectors, 175, 180
vertices, 121, 279, 282–83
visual cone, 202–4
visual proofs, 267
volume, 93

Wallis, John, 85–86, 169, 174
Weber, Heinrich, 38
Weber's tablets for Fields, 149
Weil, André, 79
Wessel, Caspar, 175, 176–77, 180
Weyl, Hermann, 48
Williams, Ted, 232
Wittgenstein, Ludwig Josef Johann, 188
Wordsworth, William, 49
The World as Will and Idea (Schopenhauer), 255

Xenophanes, 48

y-intercept, 162

Zagier, Don, 73
Zeno of Citium, 31
Zermelo, Ernst Friedrich Ferdinand, 260
zero: abstract nature of, 13–14; convergent series and, 97; correspondence and, 236; division and, 56–57; imaginary numbers and, 27; multiplication and, 7–8